Pomegranates
Ancient Roots to Modern Medicine

Medicinal and Aromatic Plants — Industrial Profiles

Individual volumes in this series provide both industry and academia with in-depth coverage of one major genus of industrial importance.

Series Edited by Dr. Roland Hardman

Pomegranates
Ancient Roots to Modern Medicine

Edited by
Navindra P. Seeram, Risa N. Schulman, and David Heber

Medicinal and Aromatic Plants — Industrial Profiles

CRC Press
Taylor & Francis Group
Boca Raton London New York

CRC Press is an imprint of the
Taylor & Francis Group, an **informa** business

A TAYLOR & FRANCIS BOOK

CRC Press
Taylor & Francis Group
6000 Broken Sound Parkway NW, Suite 300
Boca Raton, FL 33487-2742

© 2006 by Taylor and Francis Group, LLC
CRC Press is an imprint of Taylor & Francis Group, an Informa business

First issued in paperback 2019

No claim to original U.S. Government works

ISBN 13: 978-0-367-44631-4 (pbk)
ISBN 13: 978-0-8493-9812-4 (hbk)

Library of Congress Card Number 2006004494

Library of Congress Cataloging-in-Publication Data

Pomegranates : ancient roots to modern medicine / edited by Navindra P. Seeram, Risa
    N. Schulman, David Heber.
      p. ; cm. -- (Medicinal and aromatic plants--industrial profiles ; 43)
    "A CRC title."
    Includes bibliographical references and index.
    ISBN-13: 978-0-8493-9812-4 (alk. paper)
    ISBN-10: 0-8493-9812-6 (alk. paper)
    1. Pomegranate--Therapeutic use. I. Seeram, Navindra P. II. Schulman, Risa N. III.
Heber, David. IV. Series: Medicinal and aromatic plants--industrial profiles ; v. 43.
    [DNLM: 1. Plant Preparations--therapeutic use. 2. Punicaceae. 3. Phytotherapy--
methods. QV 766 P785 2006]

RM666.P798P66 2006
615'.321--dc22
                                     2006004494

Visit the Taylor & Francis Web site at
http://www.taylorandfrancis.com

and the CRC Press Web site at
http://www.crcpress.com

Preface

Wilt thou be gone? It is not yet near day.
It was the nightingale, and not the lark,
That pierced the fearful hollow of thine ear.
Nightly she sings on yon pomegranate-tree.
Believe me, love, it was the nightingale.

Romeo and Juliet, III, 5

The pomegranate is an ancient fruit that has not changed much throughout the history of man. After the discovery of agriculture about 10,000 years ago, we know that it was grown in Egypt. It was found in the Indus Valley so early that there is a word in Sanskrit for pomegranate. Indian royalty began their banquets with pomegranate, grape, and jujube.

The pomegranate is also significant in Jewish tradition. The pomegranate is said to have 613 seeds to represent the 613 commandments in the Torah. While this has not been confirmed in modern times, the image of the pomegranate was woven into the high priest's robes, and brass representations were part of the ancient Temple's pillars. The pomegranate is mentioned six times in the Song of Solomon, and even today silver-plated representations of pomegranates are used to cover the wooden handles of the Torah scrolls.

The pomegranate appeared in China during the Han and Sung dynasties, probably brought from the Middle East by traders. It was slowly adopted in medieval Europe but was known in Elizabethan England, as can be seen by the quote above from *Romeo and Juliet*. The Spanish conquistadores brought the pomegranate to America. Jesuit missionaries carried it north to California, where it now grows in abundance in that state's mild Mediterranean-like climate.

The biochemistry of the pomegranate is fascinating, and its several very different classes of compounds are discussed in detail in this book. The red/purple color of anthocyanins provides the rich color of the skin. Anthocyanins are also found in the arils (the botanical term for the part commonly consumed, including the flesh and the seed) along with traces of an astringent family of compounds known as the hydrolyzable tannins. Among these, pomegranate juice made by squeezing the whole fruit is a rich source of a large polyphenol antioxidant — an ellagitannin known as punicalagin. This molecule with a molecular weight of over 1000 Daltons does not enter the human body intact. Instead, it is broken up into ellagic acid moieties released into the bloodstream over several hours. It is also metabolized by colonic bacteria to urolithins and their conjugates, which appear in the blood and urine. The family of phytochemicals found in pomegranate juice extracts, made by squeezing the whole fruit, act together with greater potency than any of the single constituents alone, as is the case with many different fruits, vegetables, and herbs. The seeds also contain a rare lipid called punicic acid that is similar to conjugated linoleic acid but has an additional double bond.

The physiological effects of pomegranate juice constituents are remarkable in their preventive potential against two of the major chronic diseases of aging — heart disease and cancer.

First, in heart disease, the pioneering work of Michael Aviram on pomegranate polyphenols has demonstrated their ability to reduce lipid peroxidation and the progression of atherosclerosis in animal models. One clinical study demonstrated improved cardiac muscle perfusion in a group of patients with heart disease. Nobel Prize winner Lou Ignarro demonstrated marked enhancement of nitric oxide production in endothelial cells, which reveals another exciting and perhaps primary mode for the prevention of cardiovascular disease.

Second, there are significant new findings suggesting that pomegranate juice may be active against common forms of cancer. In basic laboratory studies, pomegranate juice extracts and the tannins as well as anthocyanins have significant antiproliferative and proapoptotic effects in several different types of cancer cells *in vitro* including colon cancer, prostate cancer, and head and neck cancer. In colon cancer cell lines that express cyclo-oxygenase 2 enzyme, we have demonstrated together with Bharat Aggarwal the inhibition of proliferation and Nuclear Factor kappa-B activation by pomegranate ellagitannins. In prostate cancer xenografts, pomegranate juice has been shown by Hasan Mukhtar and coworkers to reduce tumor growth. Finally, in a preliminary study of men with advanced prostate cancer and rising PSA levels performed at UCLA by Allan Pantuck, pomegranate juice reduced the rate of rise of PSA by 50% over a 1-year period. Ongoing basic and clinical research is being pursued by a number of research groups to examine the potential preventive effects of pomegranates in common forms of cancer.

Pomegranate seeds in many traditions have led to the idea that pomegranates promote fertility, but there is no evidence to that effect as yet. However, what is known of its potential phytoestrogenic effects is reviewed in some detail.

Unlocking the secrets hidden within this ancient fruit will require the application of all of our modern methods of nutritional science including nutrigenomics, proteomics, and detailed studies of subcellular signaling pathways in normal and diseased cells.

David Heber, MD, PhD, FACP, FACN
Professor of Medicine and Public Health
UCLA Center for Human Nutrition

Editors

Navindra P. Seeram, Ph.D. is assistant director of the UCLA Center for Human Nutrition and adjunct assistant professor at the David Geffen School of Medicine at the University of California, Los Angeles. His doctorate, in natural product chemistry, was obtained from the University of the West Indies and he conducted postdoctoral research at the Bioactive Natural Products and Phytoceuticals Laboratory at Michigan State University. In addition to his membership in numerous professional societies, Dr. Seeram serves on several scientific advisory boards and on editorial boards of scientific journals. His research has been widely reported in peer-reviewed journals, book chapters, and trade magazines and he is regularly invited to speak at national and international scientific conferences. He has 15 years of experience in phytochemicals and his research interests are currently focused on the evaluation of foods, spices, and herbal medicines in laboratory, animal, and human studies, for the prevention and treatment of chronic human illnesses such as cancer and heart disease.

Risa N. Schulman, Ph.D. is dedicated to the prevention of chronic disease using natural compounds found in plants and foods, and the empowerment of the individual to take control of his or her health and well-being. As the former director of research for POM Wonderful in Los Angeles, California, she built and managed an international team of researchers studying the health effects of pomegranates, making significant contributions to the development of the field, including 11 publications by the team in 3 years. At Wyeth's Solgar Vitamins, she developed plant-based nutritional supplements, patenting one formula, and managed claim substantiation for 450 products. She acted as guest curator for an exhibit on plant-based medicines in conjunction with Bristol-Myers Squibb at the New Jersey Museum of Agriculture in New Brunswick, New Jersey. Dr. Schulman's field experience includes studies of plants and ecosystems

on land and underwater in the Virgin Islands, Puerto Rico, the Bahamas, Costa Rica, Israel, and across New England. She has authored over 120 pieces designed to communicate scientific research to nonacademic groups with the American Botanical Council, currently serving as an expert reviewer. The numerous medical audiences she has spoken to include those at the Scripps Center for Integrative Medicine in La Jolla, California; the International Conference on Mechanisms of Action of Natural Products; and annual meetings of the American Dietetic Association, American Association of Family Practitioners, American Association of Naturopathic Physicians, and Pri-Med. She is the founder and former chair of Generation Health, a women's health education organization in New York City. Dr. Schulman is currently president of Schulman Scientific Consulting Inc., a business focused on planning, understanding, and communicating science, and is visiting assistant professor, Department of Clinical Nutrition, at UCLA in Los Angeles.

Originally from Trumbull, Connecticut, Dr. Schulman earned a B.S. for a double major in biology and environmental science from Tufts University, an M.E.S. in environmental science from Yale University, and a Ph.D. in plant biology from Rutgers University.

David Heber M.D., Ph.D., F.A.C.P., F.A.C.N. is director of the UCLA Center for Human Nutrition at the University of California, Los Angeles. After graduating from UCLA magna cum laude in chemistry in 1969 and from Harvard Medical School in 1973, he completed his internship at Beth Israel Hospital in Boston and his residency and fellowship training at Harbor General Hospital in Torrance, California. He earned his Ph.D. in physiology at UCLA in 1978. Dr. Heber has been on the faculty of the UCLA School of Medicine since 1978 and is currently professor of medicine and public health, founding chief of the Division of Clinical Nutrition in the Department of Medicine, and founding director of the UCLA Center for Human Nutrition.

Dr. Heber is board certified in Internal Medicine and Endocrinology and Metabolism by the American Board of Internal Medicine, and in Clinical Nutrition by the American Board of Nutrition. He directs the NCI-funded Clinical Nutrition Research Unit and the NIH Nutrition and Obesity Training Grants at UCLA. Dr. Heber is a past director of the American Board of Nutrition and past chair of the Education Committee of the American Society of Clinical Nutrition. He has written over 130 peer-reviewed scientific articles, 25 book chapters, and two professional texts: *Dietary Fat, Lipids, Hormones and Tumorigenesis,* and *Nutritional Oncology* (Academic Press, 1999; second edition due out in 2006).

Contributors

Lynn Adams
UCLA Center for Human Nutrition
David Geffen School of Medicine
University of California, Los Angeles
Los Angeles, California

Farrukh Afaq
Department of Dermatology
Medical Science Center
Madison, Wisconsin

Bharat B. Aggarwal
Cytokine Research Laboratory
Department of Experimental
  Therapeutics
M.D. Anderson Cancer Center
The University of Texas
Houston, Texas

Michael Aviram
The Lipid Research Laboratory
Rambam Medical Center
Haifa, Israel

Emily Besselink
UCLA Center for Human Nutrition
University of California, Los Angeles
Los Angeles, California

Indra D. Bhatt
Cytokine Research Laboratory
Department of Experimental
  Therapeutics
M.D. Anderson Cancer Center
The University of Texas
Houston, Texas

Juan Carlos Espín
Research Group on Quality, Safety and
  Bioactivity of Plant Food
CEBAS (CSIC)
Espinardo, Murcia, Spain

Bianca Fuhrman
The Lipid Research Laboratory
Rambam Medical Center
Haifa, Israel

Naghma Hadi
Department of Dermatology
Medical Science Center
Madison, Wisconsin

Sari Halpert
Department of Obstetrics and
  Gynecology
Columbia University
New York, New York

Diane M. Harris
UCLA Center for Human Nutrition
David Geffen School of Medicine
University of California, Los Angeles
Los Angeles, California

David Heber
UCLA Center for Human Nutrition
University of California, Los Angeles
Los Angeles, California

**Guddadarangavvanahally
  Krishanareddy Jayaprakasha**
Human Resource Development
Central Food Technological Research
  Institute
Mysore, India

Bhabani Sankar Jena
Human Resource Development
Central Food Technological Research
  Institute
Mysore, India

Adel A. Kader
Department of Plant Sciences
University of California, Davis
Davis, California

Christian G. Krueger
Department of Animal Science
University of Wisconsin–Madison
Madison, Wisconsin

John T. Leppert
Department of Urology
David Geffen School of Medicine
University of California, Los Angeles
Los Angeles, California

Arshi Malik
Department of Dermatology
Medical Science Center
University of Wisconsin
Madison, Wisconsin

Hasan Mukhtar
Department of Dermatology
Medical Science Center
University of Wisconsin
Madison, Wisconsin

Pradeep Singh Negi
Human Resource Development
Central Food Technological Research
  Institute
Mysore, India

Eliza Ng
Medical Services, Contraception
Organon International
Roseland, New Jersey

Allan J. Pantuck
Department of Urology
Center for Health Sciences
University of California, Los Angeles
Los Angeles, California

Russalind H. Ramos
Department of Obstetrics and
  Gynecology
Columbia University
New York, New York

Jess D. Reed
Department of Animal Science
University of Wisconsin–Madison
Madison, Wisconsin

Mira Rosenblat
The Lipid Research Laboratory
Rambam Medical Center
Haifa, Israel

Carolyn Schmitt
Department of Dermatology
Medical Science Center
University of Wisconsin
Madison, Wisconsin

Risa N. Schulman
Department of Clinical Nutrition
University of California, Los Angeles
Los Angeles, California
and
Schulman Scientific Consulting
Los Angeles, California

Navindra P. Seeram
UCLA Center for Human Nutrition
David Geffen School of Medicine
University of California, Los Angeles
Los Angeles, California

Shishir Shishodia
Department of Biology
Texas Southern University
Houston, Texas

David W. Still
Department of Plant Sciences
California State Polytechnic University
Pomona, California

Marva I. Sweeney-Nixon
Department of Biology and Atlantic
 Canada Network on Bioactive
 Compounds
University of Prince Edward Island
Charlottetown, Prince Edward Island,
 Canada

Deeba Syed
Department of Dermatology
Medical Science Center
University of Wisconsin
Madison, Wisconsin

Francisco A. Tomás-Barberán
Research Group on Quality, Safety and
 Bioactivity of Plant Food
CEBAS (CSIC)
Espinardo, Murcia, Spain

Jakob Vaya
Laboratory of Natural Medicinal
 Compounds
Migal-Galilee Technological Center,
 and Tel-Hai Academic College
Israel

Michelle Warren
Department of Obstetrics and
 Gynecology
Columbia University
New York, New York

Yanjun Zhang
UCLA Center for Human Nutrition
David Geffen School of Medicine
University of California, Los Angeles
Los Angeles, California

Table of Contents

SECTION 5 *Summary*

Section 1

Biochemistry

1 Pomegranate Phytochemicals

Navindra P. Seeram, Yanjun Zhang, Jess D. Reed,
Christian G. Krueger, and Jakob Vaya

CONTENTS

1.1 INTRODUCTION

Phytochemicals are secondary plant metabolites that have human health benefits but
are not considered essential nutrients. Examples of essential nutrients are proteins,

fats, carbohydrates, vitamins, and minerals. Therefore phytochemicals are often referred to as "nonnutritive" compounds thought to be produced by plants as means of protection against such dangers as harmful ultraviolet (UV) radiation, pathogens, and herbivorous predators. Phytochemicals possess a wide range of structural variations, which impart unique chemical and biological properties to their classes and subclasses.

The consumption of a plant-based or phytochemical-rich diet has been associated with a reduced risk of chronic human illnesses such as certain types of cancers, inflammation, and cardiovascular and neurodegenerative diseases. Therefore the chemistry (isolation, identification, structural elucidation, and characterization) and biology (*in vitro* and *in vivo* bioactivities, mechanisms of action, absorption, distribution, metabolism, and excretion) of phytochemicals are of utmost importance for evaluation of their potential health benefits to humans. This chapter reviews the different types and classes of phytochemicals that have been identified from the pomegranate (*Punica granatum* L.) fruit (arils, rind, pith, pericarp, and seeds) and tree (bark, stems, leaves, and roots) with a focus on the dietary phytochemicals, in particular those found in the edible parts (fruit, peel, and seeds). Phytochemical methods of analyses as well as the roles that pomegranate phytochemicals play in bioactivity are also discussed.

1.2   POMEGRANATE PHYTOCHEMICALS

The different types of phytochemicals that have been identified from various parts of the pomegranate tree and from pomegranate fruits and seeds are listed in Table 1.1, items 1 to 31, and Figure 1.1a through Figure 1.1m show their chemical structures. The major class of pomegranate phytochemicals is the polyphenols (phenolic rings bearing multiple hydroxyl groups) that predominate in the fruit. Pomegranate polyphenols include flavonoids (flavonols, flavanols, and anthocyanins), condensed tannins (proanthocyanidins), and hydrolyzable tannins (ellagitannins and gallotannins). Other phytochemicals identified from the pomegranate are organic and phenolic acids, sterols and triterpenoids, fatty acids, triglycerides, and alkaloids.

The major source of dietary pomegranate phytochemicals is the fruit. Pomegranates are popularly consumed as fresh fruit, as beverages (e.g., juices and wines), as food products (e.g., jams and jellies), and as extracts wherein they are used as botanical ingredients in herbal medicines and dietary supplements. Commercial pomegranate juice (PJ) is obtained by a hydrostatic pressing process of whole fruits whereby two predominant types of polyphenolic compounds are extracted into PJ: flavonoids and hydrolyzable tannins (HTs).[32] The flavonoids include flavonols such as luteolin, quercetin, and kaempferol found in the peel extract[17] and anthocyanins found in the arils.[14,15] Anthocyanins are the water-soluble pigments responsible for the bright red color of PJ. Pomegranate anthocyanins include pelargonidin-3-glucoside, cyanidin-3-glucoside, delphinidin-3-glucoside, pelargonidin 3,5-diglucoside, cyanidin 3,5-diglucoside, and delphinidin 3,5-diglucoside[14,15] (Table 1.1).

HTs are found in the peels (rind, husk, or pericarp), membranes, and piths of the fruit.[33] HTs are the predominant polyphenols found in PJ and account for 92%

TABLE 1.1
Phytochemicals Identified from Pomegranates

ID	Name	Formula	MW	Plant Part	Ref.
	Ellagitannins and Gallotannins				
1	2,3-(S)-HHDP-D-glucose[a]	$C_{20}H_{18}O_{14}$	482.35	Bark, peel	1
2	Castalagin	$C_{41}H_{26}O_{26}$	934.63	Bark	2
3	Casuariin	$C_{34}H_{24}O_{22}$	784.54	Bark	2
4	Casuarinin	$C_{41}H_{28}O_{26}$	936.65	Bark, pericarp	2, 3
5	Corilagin	$C_{27}H_{22}O_{18}$	634.45	Fruit, leaves, pericarp	3, 4, 5
6	Cyclic 2,4:3,6-bis(4,4′,5,5′,6,6′-hexahydroxy[1,1′-biphenyl]-2,2′-dicarboxylate) 1-(3,4,5-trihydroxybenzoate) b-D-Glucose	$C_{41}H_{28}O_{26}$	936.65	Leaves	6
7	Granatin A	$C_{34}H_{24}O_{23}$	800.54	Pericarp	3
8	Granatin B	$C_{34}H_{28}O_{27}$	952.64	Peel	3, 4
9	Pedunculagin	$C_{34}H_{24}O_{22}$	784.52	Bark, pericarp	1, 3
10	Punicacortein A	$C_{27}H_{22}O_{18}$	634.45	Bark	2
11	Punicacortein B	$C_{27}H_{22}O_{18}$	634.45	Bark	2
12	Punicafolin	$C_{41}H_{30}O_{26}$	938.66	Leaves	7
13	Punigluconin	$C_{34}H_{26}O_{23}$	802.56	Bark	2
14	Strictinin	$C_{27}H_{22}O_{18}$	634.45	Leaves	7
15	Tellimagrandin I	$C_{34}H_{26}O_{22}$	786.56	Leaves, pericarp	3, 8
16	Tercatain	$C_{34}H_{26}O_{22}$	786.56	Leaves	8
17	2-O-galloyl-4,6(S,S) gallagoyl-D-glucose	$C_{41}H_{26}O_{26}$	934.63	Bark	1
18	5-O-galloyl-punicacortein D	$C_{54}H_{34}O_{34}$	1222.8	Leaves	9
19	Punicacortein C	$C_{47}H_{26}O_{30}$	1070.7	Bark	2
20	Punicacortein D	$C_{47}H_{26}O_{30}$	1070.7	Bark, heartwood	2
21	Punicalin	$C_{34}H_{22}O_{22}$	782.53	Bark, pericarp	1, 3
22	Punicalagin	$C_{48}H_{28}O_{30}$	1084.7	Bark, pericarp, peel	1, 2, 3
23	Terminalin/gallayldilacton	$C_{28}H_{20}O_{16}$	602.37	Pericarp	3
	Ellagic Acid Derivatives				
24	Ellagic acid	$C_{14}H_6O_8$	302.19	Fruit, pericarp, bark	3, 10
25	Ellagic acid, 3,3′-di-O-methyl	$C_{16}H_{10}O_8$	330.25	Seed	11
26	Ellagic acid, 3,3′, 4′-tri-O-methyl	$C_{17}H_{12}O_8$	344.27	Seed	11
27	Ellagic acid, 3′-O-methyl-3, 4-methylene	$C_{16}H_8O_8$	328.23	Heartwood	5
28	Eschweilenol C	$C_{20}H_{16}O_{12}$	448.33	Heartwood	5
29	Diellagic acid rhamnosyl(1-4) glucoside	$C_{40}H_{30}O_{24}$	894.65	Heartwood	9

(continued)

TABLE 1.1 (continued)
Phytochemicals Identified from Pomegranates

ID	Name	Formula	MW	Plant Part	Ref.
		Catechin and Procyanidins			
30	(-)-Catechin	$C_{15}H_{14}O_6$	290.27	Juice	12
31	Catechin-(4,8)-gallocatechin	$C_{30}H_{26}O_{13}$	594.52	Peel	13
32	Gallocatechin	$C_{15}H_{14}O_7$	306.27	Peel	13
33	Gallocatechin-(4,8)-catechin	$C_{30}H_{26}O_{13}$	594.52	Peel	13
34	Gallocatechin-(4,8)-gallocatechin	$C_{30}H_{26}O_{14}$	610.52	Peel	13
35	Procyanidin B1	$C_{30}H_{26}O_{12}$	578.52	Juice	12
36	Procyanidin B2	$C_{30}H_{26}O_{12}$	578.52	Juice	12
		Anthocyanins and Anthocyanidins			
37	Cyanidin	$C_{15}H_{11}O_6$	287.24	Juice	4, 14
38	Cyanidin-3-glucoside	$C_{21}H_{21}O_{11}$	449.38	Juice	4, 14, 15
39	Cyanidin-3,5-diglucoside	$C_{27}H_{31}O_{16}$	611.52	Juice	4, 14, 15
40	Cyanidin-3-rutinoside	$C_{27}H_{31}O_{15}$	595.53	Juice	12
41	Delphinidin	$C_{15}H_{11}O_7$	303.24	Juice	4, 14
42	Delphinidin-3-glucoside	$C_{21}H_{21}O_{12}$	465.38	Juice	4, 14, 15
43	Delphinidin 3, 5-diglucoside	$C_{27}H_{31}O_{17}$	627.52	Juice	4, 14, 15
44	Pelargonidin 3-glucoside	$C_{21}H_{21}O_{10}$	433.38	Juice	4, 14
45	Pelargonidin 3,5-diglucoside	$C_{27}H_{31}O_{15}$	595.53	Juice	4, 14
		Flavonols			
46	Apigenin-4′-O-β-D-glucoside	$C_{21}H_{20}O_{11}$	448.32	Leaves	16
47	Kaempferol	$C_{15}H_{10}O_6$	286.24	Peel, fruit	17
48	Luteolin	$C_{15}H_{10}O_6$	286.24	Peel, fruit	17
49	Luteolin-3′-O-β-D-glucoside	$C_{21}H_{20}O_{10}$	432.11	Leaves	16
50	Luteolin-4′-O-β-D-glucoside	$C_{21}H_{20}O_{10}$	432.11	Leaves	16
51	Luteolin-3′-O-β-D-Xyloside	$C_{20}H_{18}O_{10}$	418.09	Leaves	16
52	Myricetin	$C_{15}H_{10}O_8$	318.04	Fruit	18
53	Quercetin	$C_{15}H_{10}O_7$	302.04	Peel, fruit	4, 18, 19
54	Quercimeritrin	$C_{21}H_{20}O_{12}$	464.38	Fruit	4
55	Quercetin-3-O-rutinoside	$C_{27}H_{30}O_{16}$	610.52	Fruit	12
56	Quercetin-3,4′-dimethyl ether 7-O-α-L-arabinofuranosyl-(1-6)-β-D-glucoside	$C_{28}H_{32}O_{16}$	624.54	Bark, peel	19
57	Eriodictyol-7-O-α-L-arabinofuranosyl (1-6)-β-D-glucoside	$C_{26}H_{30}O_{15}$	582.51	Leaves	20
58	Naringenin 4′-methylether 7-O-α-L-arabinofuranosyl (1-6)-β-D-glucoside	$C_{27}H_{32}O_{14}$	580.53	Leaves	20
		Organic Acids			
59	Caffeic acid	$C_9H_8O_4$	180.16	Juice	12
60	Chlorogenic acid	$C_{16}H_{18}O_9$	345.31	Juice	12

TABLE 1.1 (continued)
Phytochemicals Identified from Pomegranates

ID	Name	Formula	MW	Plant Part	Ref.
61	Cinnamic acid	$C_9H_8O_2$	148.16	Juice	12
62	Citric acid	$C_6H_8O_7$	192.12	Juice	21
63	o-Coumaric acid	$C_9H_8O_3$	164.16	Juice	12
64	p-Coumaric acid	$C_9H_8O_3$	164.16	Juice	12
65	Ferulic acid	$C_{10}H_{10}O_4$	194.18	Juice	12
66	Gallic acid	$C_7H_6O_5$	170.12	Juice	12
67	L-Malic acid	$C_4H_6O_5$	134.09	Juice	21
68	Oxalic acid	$C_2H_2O_4$	90.03	Juice	21
69	Protocatechuic acid	$C_7H_6O_4$	154.12	Juice	21
70	Quinic acid	$C_7H_{12}O_6$	192.17	Juice	12
71	Succinic acid	$C_4H_6O_4$	118.09	Juice	21
72	Tartaric acid	$C_4H_6O_6$	150.09	Juice	21
	Simple Gallyol Derivatives				
73	Brevifolin	$C_{12}H_8O_6$	248.19	Leaves	10
74	Brevifolin carboxylic acid	$C_{13}H_8O_8$	292.2	Leaves	10
75	Brevifolin carboxylic acid-10-monosulphate	$C_{13}H_7KO_{10}S$	394.25	Leaves	8
76	1,2,3-Tri-O-galloyl-β-D-glucose	$C_{27}H_{24}O_{18}$	636.47	Leaves	10
77	1,2,4-Tri-O-galloyl-β-D-glucose	$C_{27}H_{24}O_{18}$	636.47	Leaves	10
78	1,2,6-Tri-O-galloyl-β-D-glucose	$C_{27}H_{24}O_{18}$	636.47	Leaves	10
79	1,4,6-Tri-O-galloyl-β-D-glucose	$C_{27}H_{24}O_{18}$	636.47	Leaves	10
80	1,3,4-Tri-O-galloyl-β-D-glucose	$C_{27}H_{24}O_{18}$	636.47	Leaves	8
81	1,2, 4, 6-Tetra-O-galloyl-β-D-glucose	$C_{34}H_{28}O_{22}$	788.57	Leaves	10
82	1,2,3,4, 6-Pent-O-galloyl-β-D-glucose	$C_{41}H_{32}O_{26}$	940.68	Leaves	10
83	Methyl gallate	$C_8H_8O_5$	184.15	Heartwood	5
84	3,4,8,9,10-pentahydroxy-dibenzo[b,d]pyran-6-one	$C_{13}H_8O_7$	276.20	Leaves	10
	Fatty Acids and Triglycerides				
85	Eicosenoic acid	$C_{20}H_{40}O_2$	312.53	Seed oil	22
86	Linoleic acid	$C_{18}H_{32}O_2$	280.45	Seed oil	2
87	Linolenic acid	$C_{18}H_{30}O_2$	278.43	Seed oil	22
88	Oleic acid	$C_{18}H_{34}O_2$	282.46	Seed oil	22
89	Palmitic acid	$C_{16}H_{32}O_2$	256.42	Seed oil	22
90	Punicic acid	$C_{18}H_{30}O_2$	278.43	Seed oil	22
91	Stearic acid	$C_{18}H_{36}O_2$	284.48	Seed oil	22
92	Tri-O-punicylglycerol	$C_{57}H_{92}O_6$	873.34	Seeds	23
93	Di-O-punicyl-O-octadeca-8Z-11Z-13E-trienylglycerol	$C_{57}H_{92}O_6$	873.34	Seeds	23
94	1-O-trans, cis, trans, octadecatrienol glycerol	$C_{21}H_{36}O_4$	352.51	Seed oil	24

(continued)

TABLE 1.1 (continued)
Phytochemicals Identified from Pomegranates

ID	Name	Formula	MW	Plant Part	Ref.
95	1-*O*-isopentyl-3-*O*-octadec-2-enoyl glycerol	$C_{26}H_{50}O_4$	426.67	Seed oil	24

Sterols and Terpenoids

ID	Name	Formula	MW	Plant Part	Ref.
96	Asiatic acid	$C_{30}H_{48}O_5$	488.7	Seed	25
97	Betulinic acid	$C_{30}H_{48}O_3$	456.70	Seed	26
98	Cholesterol	$C_{27}H_{46}O$	386.65	Seed oil	27
99	Daucosterol	$C_{35}H_{60}O_6$	576.85	Seed	11
100	Estrone	$C_{18}H_{22}O_2$	270.37	Seed oil	27
101	Estradiol	$C_{18}H_{24}O_2$	272.38	Seed oil	27
102	Estriol	$C_{18}H_{24}O_3$	288.38	Seed oil	27
103	Friedooleanan-3-one	$C_{30}H_{50}O$	426.72	Bark	28
104	β-Sitosterol	$C_{29}H_{50}O$	414.71	Seed oil, leaves, stem	27, 29
105	Stigmasterol	$C_{29}H_{48}O$	412.69	Seed oil	27
106	Testosterone	$C_{19}H_{28}O_2$	288.42	Seed oil	27
107	Ursolic acid	$C_{30}H_{48}O_3$	456.70	Seed	29

Alkaloids

ID	Name	Formula	MW	Plant Part	Ref.
108	Hygrine	$C_8H_{15}NO$	141.21	Root bark	30
109	Norhygrine	$C_7H_{13}NO$	127.18	Root bark	30
110	Pelletierine	$C_8H_{15}NO$	141.21	Bark	30
111	N-methyl pelletierine	$C_9H_{17}NO$	155.24	Bark	30
112	Sedridine	$C_8H_{17}NO$	143.23	Bark	30
113	Pseudopelletierine	$C_9H_{15}NO$	153.22	Bark	30
114	Nor-pseudopelletierine	$C_8H_{13}NO$	139.19	Bark	30
115	2,3,4,5-tetrahydro-6-propenyl-pyridine	$C_8H_{13}N$	123.20	Bark	30
116	3,4,5,6-tetrahydro-a-methyl-2-pyridine ethanol	$C_8H_{15}NO$	141.21	Bark	30
117	1-(2,5-dyihydroxy-phenyl)-pyridium chloride	$C_{11}H_{10}ClNO_2$	223.66	Leaves	16

Other Compounds

ID	Name	Formula	MW	Plant Part	Ref.
118	Coniferyl 9-O-[β-D-apiofuranosyl-(1-6)]-O-β-D-glucopyranoside	$C_{21}H_{30}O_{12}$	474.46	Seed	11
119	Sinapyl 9-O-[β-D-apiofuranosyl-(1-6)]-O-β-D-glucopyranoside	$C_{22}H_{32}O_{13}$	504.48	Seed	11
120	Phenylethyl rutinoside	$C_{20}H_{30}O_{10}$	430.45	Seed	11
121	Icariside D1	$C_{19}H_{28}O_{10}$	416.42	Seed	11
122	Mannitol	$C_6H_{14}O_6$	182.17	Bark	31

[a] HHDP = hexahydroxydiphenoyl.

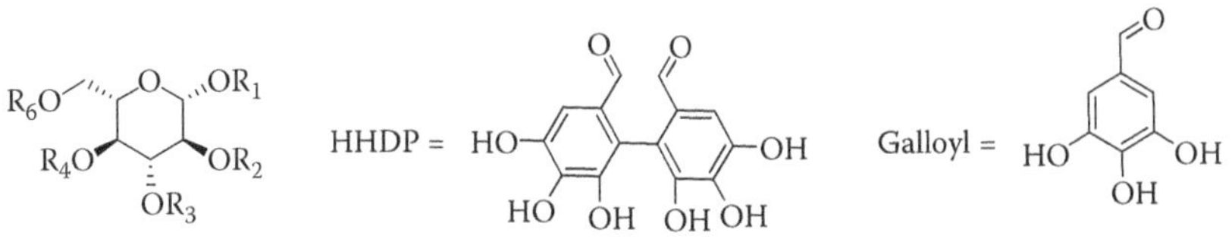

		R1	R2	R3	R4	R6
1		H		^{2}HHDP^3	H	H
5		Galloyl	H	^{3}HHDP^6	H	
6		Galloyl	^{2}HHDP^4			^{3}HHDP^6
9		H	^{2}HHDP^3			^{4}HHDP^6
12		Galloyl	Galloyl	^{3}HHDP^6	Galloyl	
14		Galloyl				^{4}HHDP^6
15			Galloyl	Galloyl		^{4}HHDP^6
16		Galloyl		Galloyl		^{4}HHDP^6

(a)

FIGURE 1.1 Chemical structures of pomegranate phytochemicals; a = ellagitannins and gallotannins; b = ellagic acid derivatives; c = catechins and procyanidins; d = anthocyanins and anthocyanidins; e = organic and phenolic acids; f = simple galloyl derivatives; g = fatty acids and triglycerides; h = sterols and triterpenoids; i = alkaloids; j = other compounds.

of its antioxidant activity.[32] The predominant pomegranate HT is punicalagin, which is responsible for about half of the total antioxidant capacity of the juice. The soluble polyphenol content in PJ has been reported to vary within the limits of 0.2 to 1.0%.[34] HTs are gallic acid and ellagic acid esters of core molecules that consist of polyols such as sugars. HTs are susceptible to enzymatic and non-enzymatic hydrolysis and are further classified according to the products of hydrolysis: gallotannins yield gallic acid and glucose while ellagitannins yield ellagic acid and glucose.[35] This combination of monomers leads to large structural heterogeneity. For instance, pomegranate leaves, bark, and fruit contain more than 18 individual HT structures.[1,2,7] These structures are classified into gallotannins, ellagitannins (ellagic acid esters of D-glucose with one or more galloyl substitutions) and the more unique gallagoyl esters such as punicalagin and punicalin (Figure 1.1a to Figure 1.1d).

Recently, there has been an increase in the use of pomegranate fruit extracts as botanical ingredients in herbal medicines and dietary supplements. In these extracts, both polyphenol and fatty-acid constituents from various parts of the fruit such as its arils and juice, peels, pericarp and pith, and seeds, may be present. The seeds of pomegranates have been well investigated. For example, pomegranate seed oil

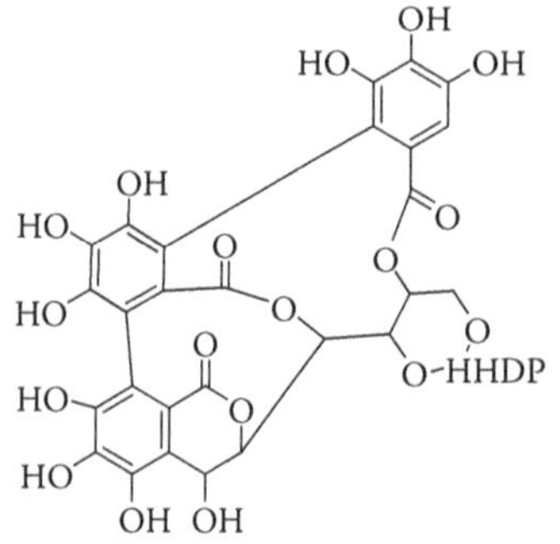

2

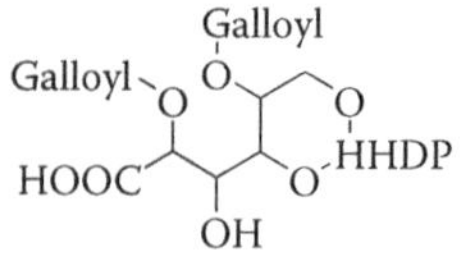

13

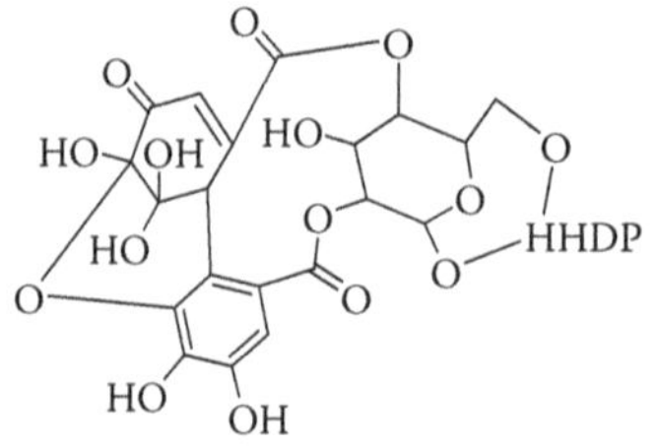

(b)

7

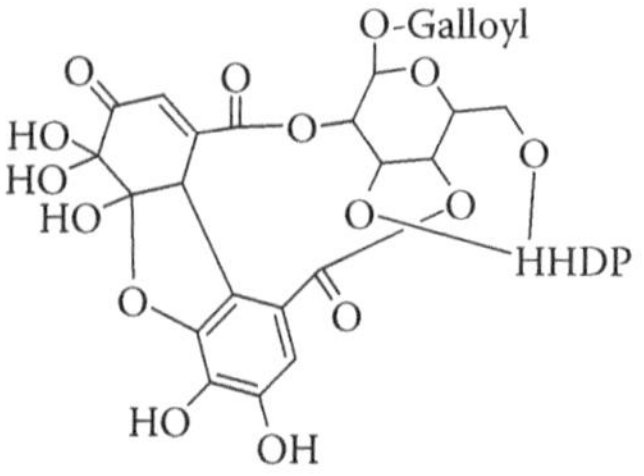

8

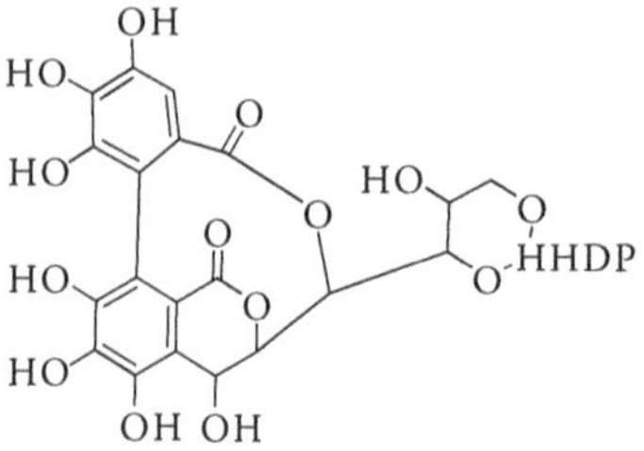

3

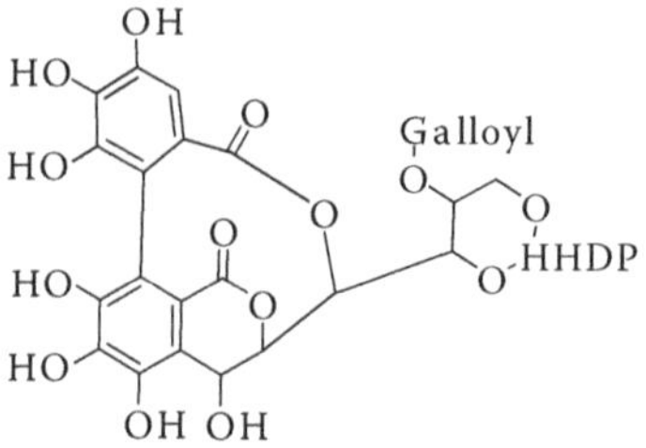

4

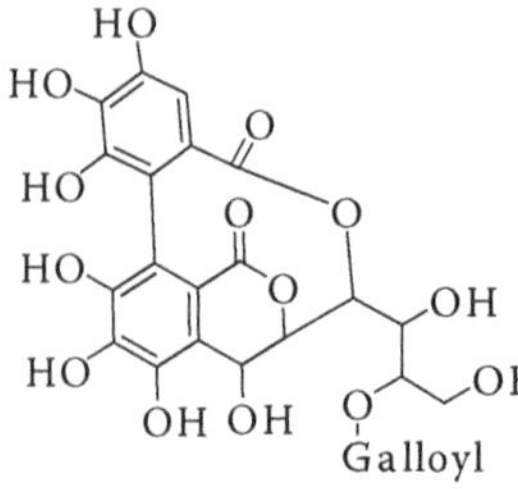

10

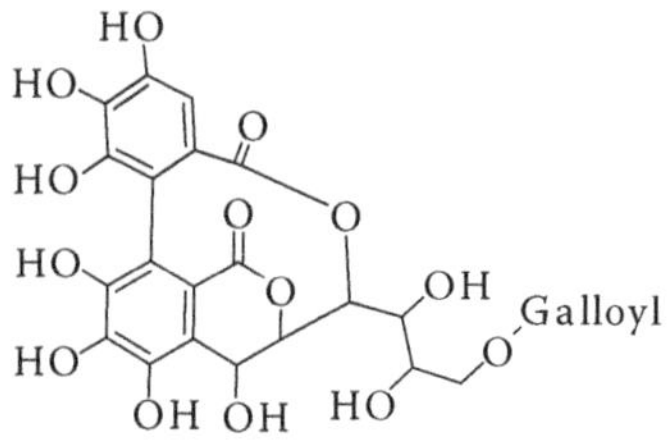

11

(c)

FIGURE 1.1 (continued)

FIGURE 1.1 (continued)

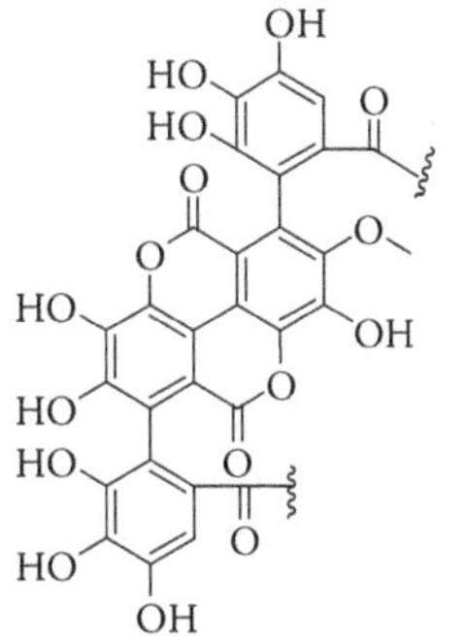

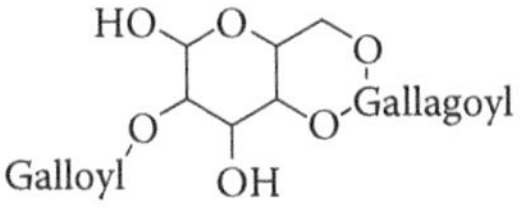

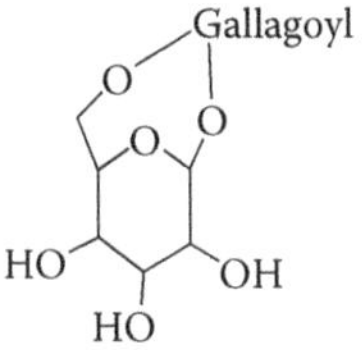

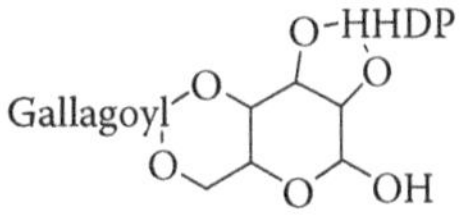

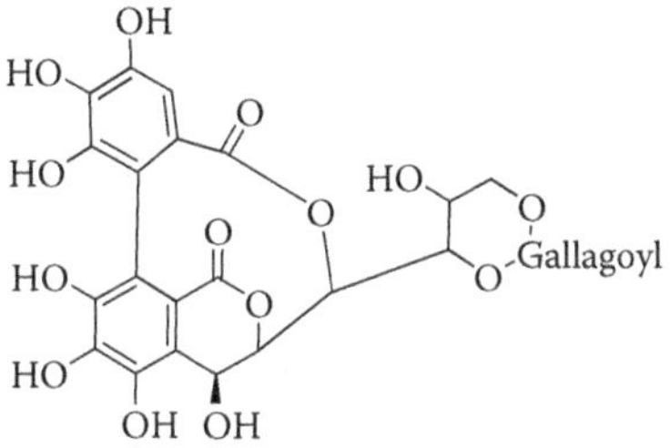

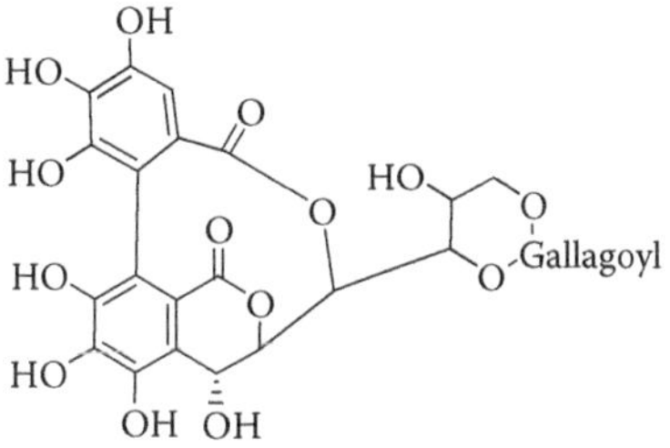

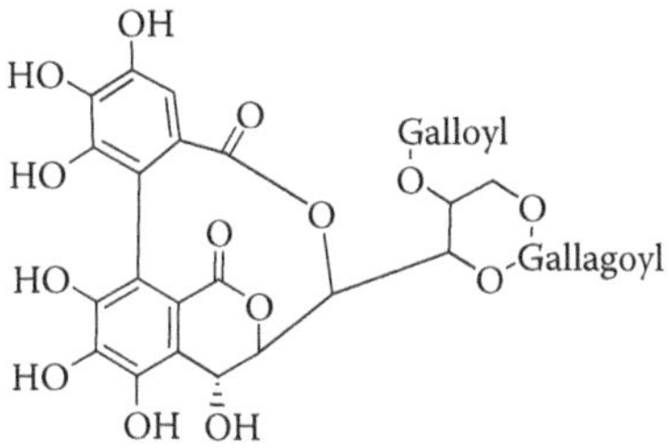

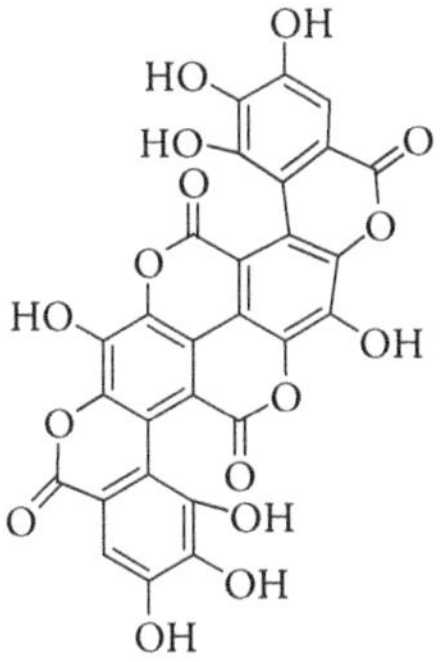

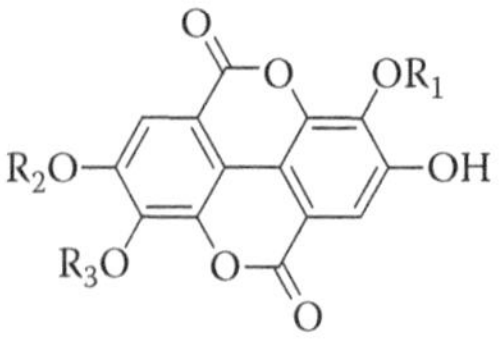

	R1	R2	R3
24	H	H	H
25	Me	H	Me
26	Me	Me	Me
27	Me	-CH$_2$-	
28	H	Rhamnose	H

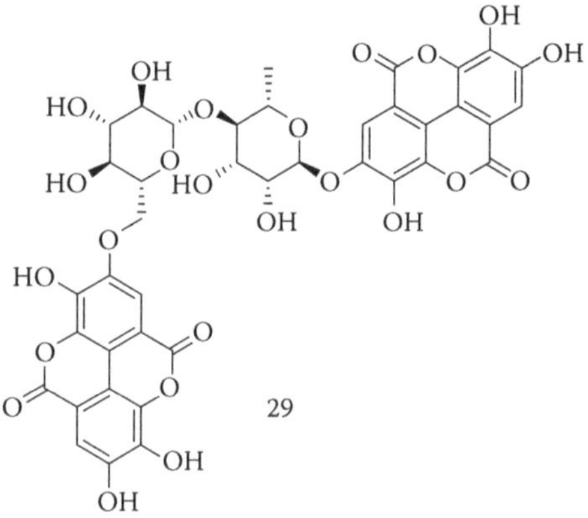

(e)

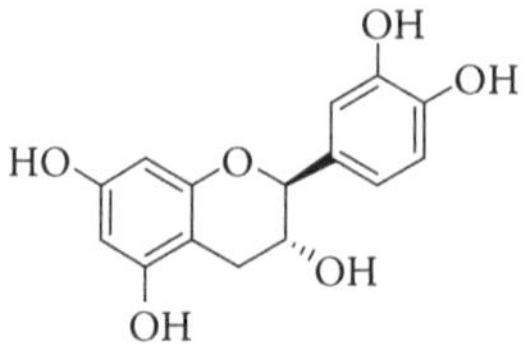

31 Catechin-(4-8)-gallocatechin

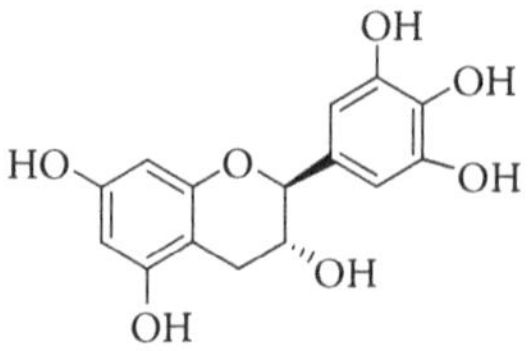

33 Gallocatechin (4-8)-catechin

34 Gallocatechin-(4-8)-gallocatechin

35 Epicatechin-(4-8)-catechin

36 Epicatechin-(4-8)-epicatechin

(f)

FIGURE 1.1 (continued)

contains levels of >60% of punicic acid, a cis-9, trans-11, cis-13 conjugated linolenic acid.[36] Other fatty acids found in pomegranate seed oil include linoleic, oleic, palmitic, stearic, arachidic, and palmitoleic acid.[37] In addition, there have been reports of sex hormones such as estrone and coumestrol in pomegranate seeds.[37–39] However, a recent paper did not detect the presence of the steroid hormones estone, estradiol, and testosterone in pomegranate seeds, juice, and commercial preparations.

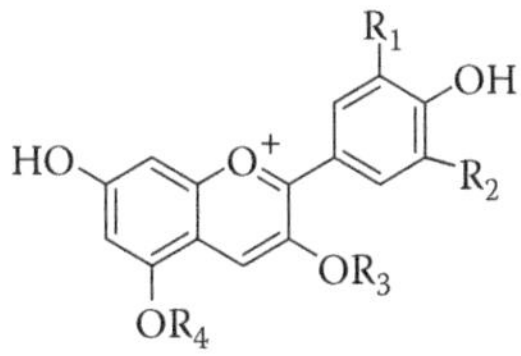

	R1	R2	R3	R4
37	OH	H	H	H
38	OH	H	Glucose	H
39	OH	H	Glucose	Glucose
40	OH	H	Rutinose	H
41	OH	OH	H	H
42	OH	OH	Glucose	H
43	OH	OH	Glucose	Glucose
44	H	H	Glucose	H
45	H	H	Glucose	Glucose

(g)

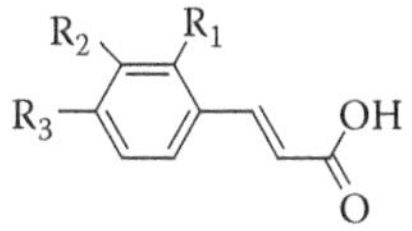

	R1	R2	R3
59	H	OH	OH
61	H	H	H
63	OH	H	H
64	H	H	OH
65	H	OMe	OH

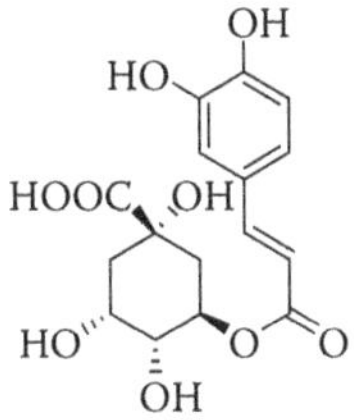

60 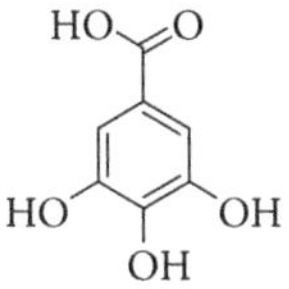   66 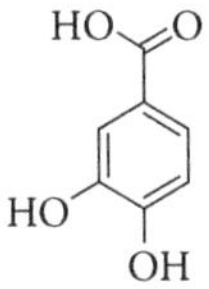   67 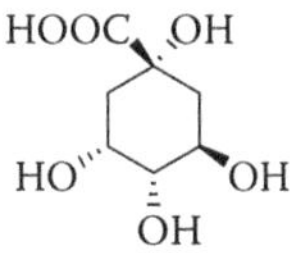   68

(h)

FIGURE 1.1 (continued)

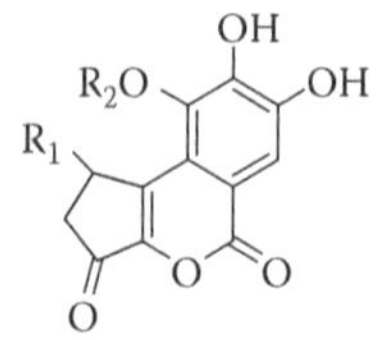

73 R_1 = H R_2 = H   83   84
74 R_1 = COOH R_2 = H
75 R_1 = COOH R_2 = SO_2K

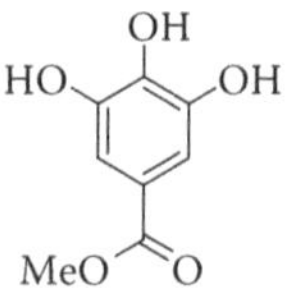

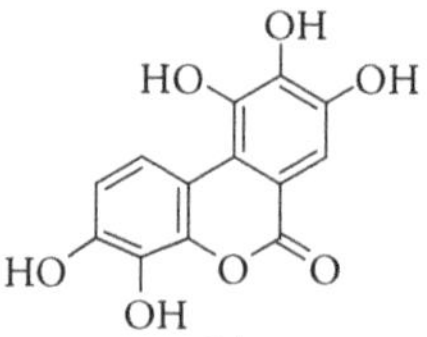

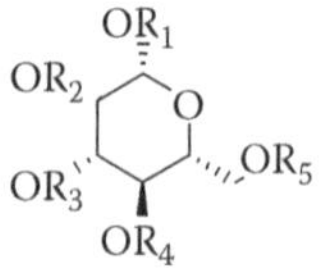

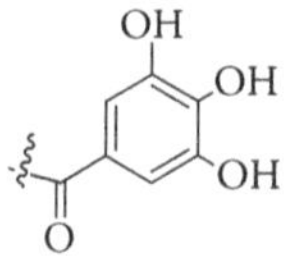

	R1	R2	R3	R4	R5
76	G	G	G	H	H
77	G	G	H	G	H
78	G	G	H	H	G
79	G	H	H	G	G
80	G	H	G	G	H
81	G	G	H	G	G
82	G	G	G	G	G

(i)

FIGURE 1.1 (continued)

1.3 METHODS FOR SEPARATION, IDENTIFICATION, AND ANALYSES OF POMEGRANATE PHYTOCHEMICALS

1.3.1 CHROMATOGRAPHY

Absorption and adsorption chromatography using resins such as C-18, Sephadex-LH20, and Amberlite XAD are widely used for the isolation of pomegranate phytochemicals, in particular its polyphenols.[1,2,7,33] Chromatographic methods using preparative- and semipreparative-scale high-performance liquid chromatography (HPLC) and medium-pressure liquid chromatography (MPLC) are also used for producing semipurified fractions for mass spectral analysis of crude extracts that would otherwise discriminate against high-molecular-weight (MW) components. To overcome this problem, absorption chromatography has also been used to produce more homogeneous fractions of narrower MW distributions prior to analysis by mass spectrometry.[41–45] This approach produces spectra of higher resolution of individual

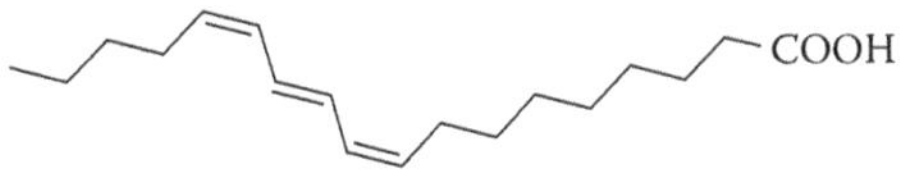

90

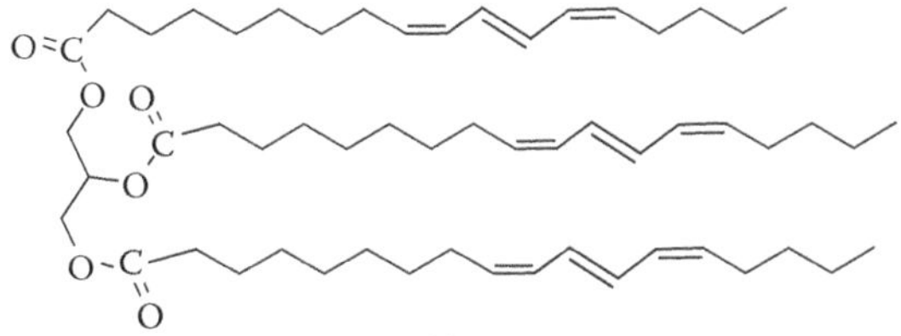

92

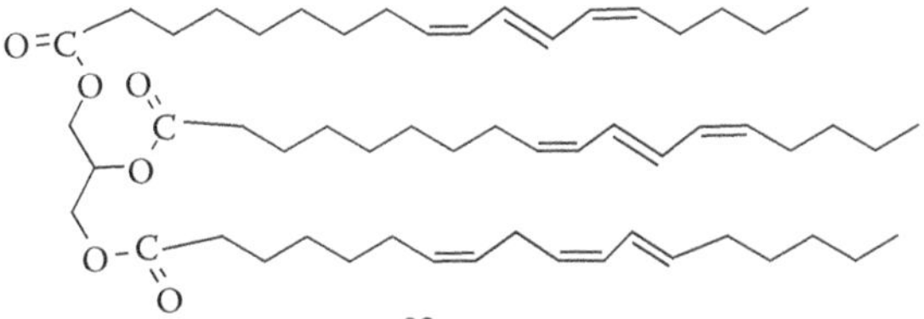

93

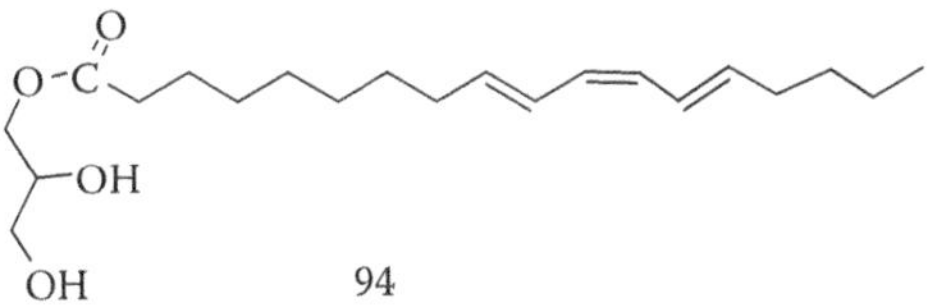

94

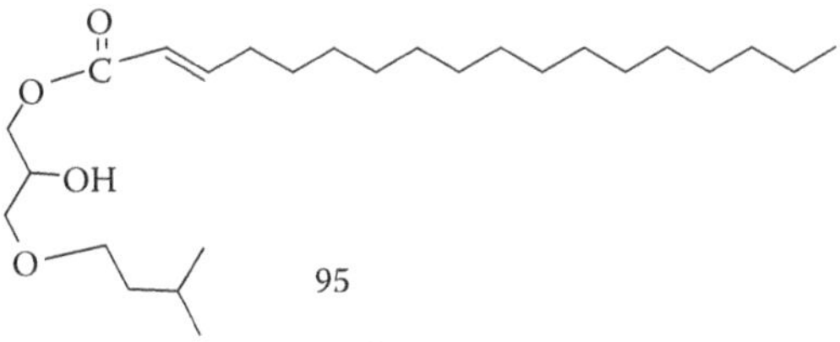

95

(j)

FIGURE 1.1 (continued)

compounds and allows for better interpretation of the structural variation of pomegranate oligomers. The mass spectrometric methods are discussed below (Section 1.3.2).

1.3.2 High-Performance Liquid Chromatography (HPLC), Gas Chromatography, and Mass Spectrometry (MS)

There are immense structural variations within polyphenols extracted from fruits, juices, and other parts of pomegranate. Underestimation of the structural complexity and range in degree of polymerization present in polyphenols negatively impacts the interpretation of the relationship between the structure of oligomeric polyphenols

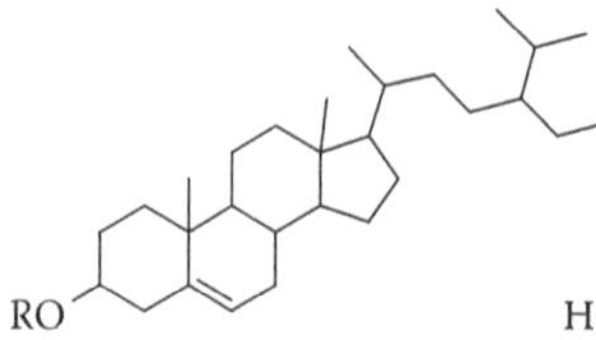

104 R = H
99   R = glucose

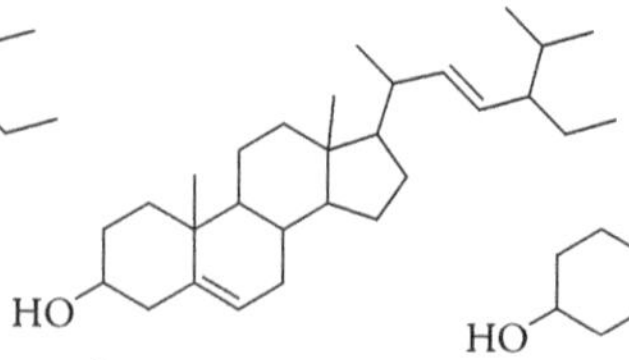

105

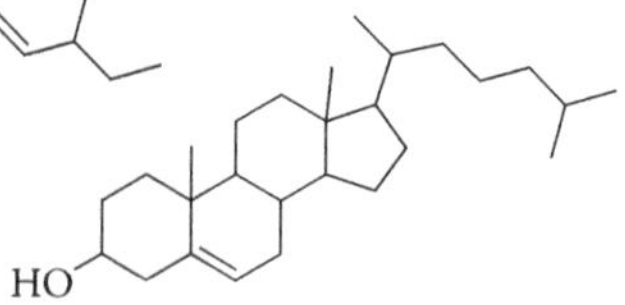

98

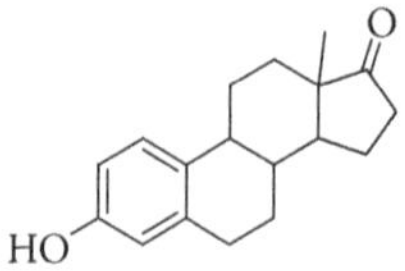

100

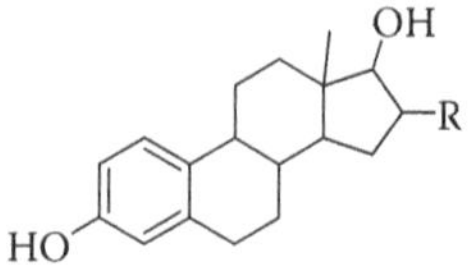

101 R = H
102 R = OH

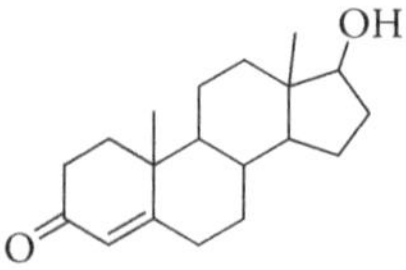

106

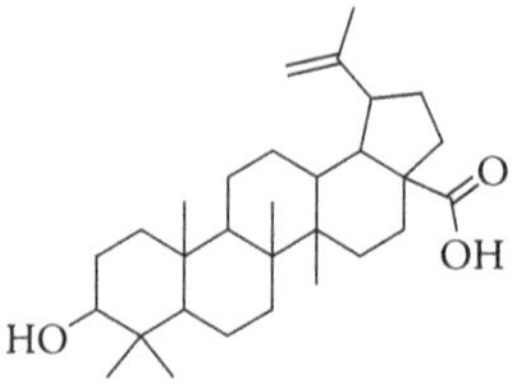

97

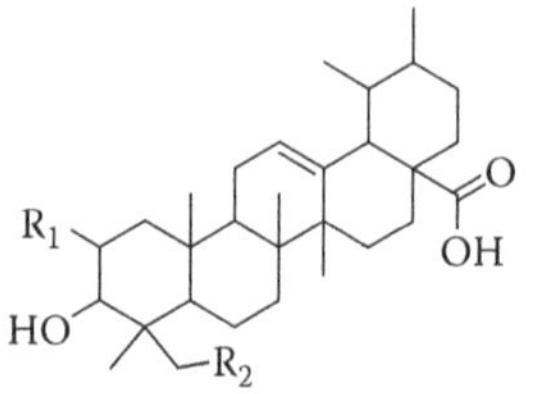

96   R_1 = OH, R_2 = OH
107 R_1 = H, R_2 = H

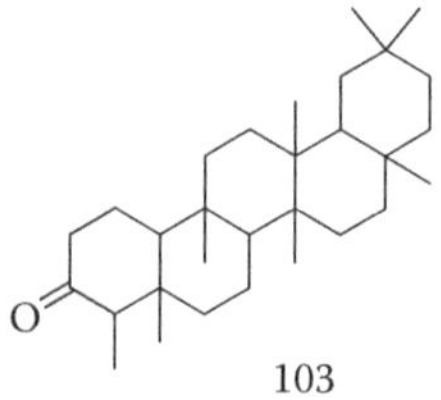

103

(k)

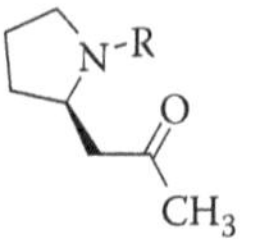

108 R = Me
109 R = H

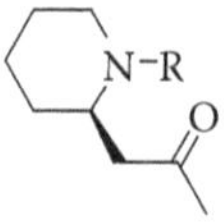

110 R = H
111 R = Me

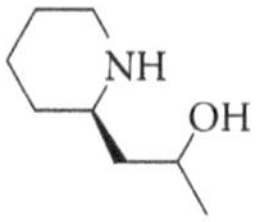

112

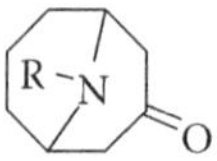

113 R = Me
114 R = H

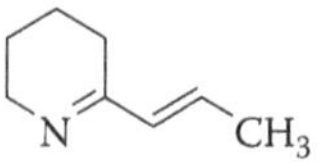

115

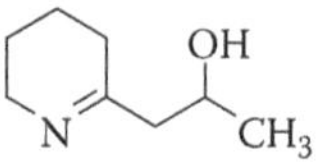

116

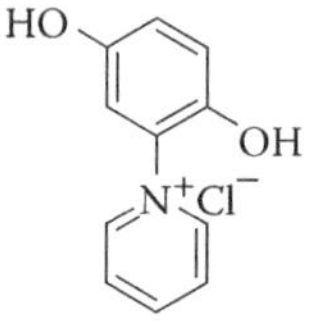

117

(l)

FIGURE 1.1 (continued)

and their biological function. High-performance liquid chromatography (HPLC) coupled with ultraviolet (UV), fluorescence, and electrochemical detectors (ECDs) have been used for the identification of pomegranate polyphenols. However, these

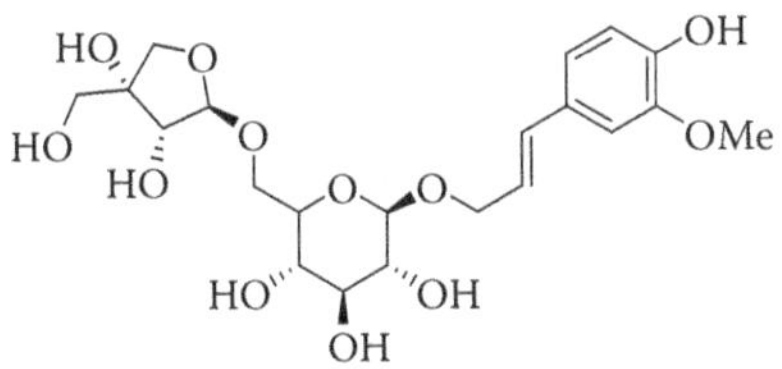

(m)

FIGURE 1.1 (continued)

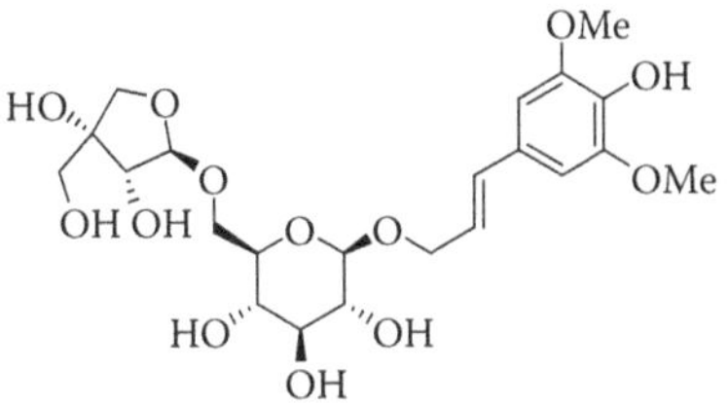

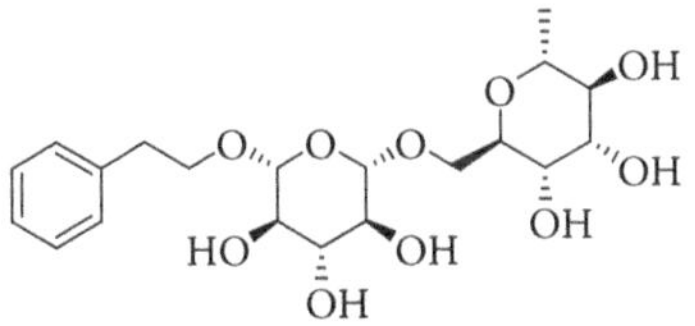

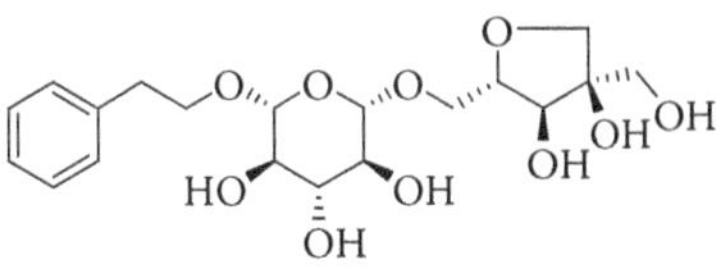

methods have severe limitations in the absence of authentic standards for particular phytochemical(s). Recent advances in the field of mass spectrometry detection provide a new tool for the investigation of oligomeric polyphenols. Tentative structural assignments of oligomeric polyphenols are made possible by applying predictive equations to interpret complex mass spectra.[44] Predictive equations are based on known monomeric ellagitannin structures and the assumption that the structural diversity seen in ellagitannins can be extrapolated to higher degrees of polymerization by linking two or more core glucose molecules. Mass spectrometric methods that have been used for the identification of pomegranate phytochemicals are as follows.

1.3.2.1 MALDI-TOF Mass Spectrometry

Reports on the application of matrix-assisted laser desorption/ionization time-of-flight mass spectrometry (MALDI-TOF MS) for the analysis of oligomeric HTs have appeared within the last 3 years. More recently, a series of oligomeric HTs was detected in pomegranate aril extracts using MALDI-TOF MS.[46] This oligomeric series consisted of at least 22 HTs with masses between 500 and 4058 amu. Masses less than 1500 amu corresponded to known pomegranate ellagitannin and gallotannin structures, such as punicalagin as described previously.[1,2,7] The higher masses correspond to structures of oligomeric ellagitannins in which two or more core glucose units are cross-linked by dihydrodigalloyl or valoneoyl units (Table 1.2). Tentative structural assignments of compounds are based on the following substitutions: glucose (glucosyl), gallagic acid (gallagoyl), hexahydroxydiphenoyl (HHDP), gallic acid (galloyl), and dehydrodigalloyl (DHDG). This was the first time that these higher oligomers have been detected in pomegranates,[46] although they are known to occur in other plants.[47]

TABLE 1.2
MALDI-TOF Mass Spectrometry of the Ellagitannins and Other Hydrolyzable Tannins in Pomegranate Fruit Extract

Ellagitannins and Other Hydrolyzable Tannins	Expected Mass [M + Cs]$^+$	Observed Mass [M + Cs]$^+$	Glucosyl	Gallagoyl	Hexahydroxybiphenoyl	Galloyl	Dehydrodigalloyl
Punicalin	914.4	914.5	1	1	0	0	0
Pedunculagin	916.4	916.5	1	0	2	0	0
	918.5	918.6	1	0	1	2	0
Punicalagin	1216.6	1217.1	1	1	1	0	0
	1218.7	1219.1	1	1	0	2	0
Dimers	1550.9	1551.1	2	0	2	1	1
	1701.0	1701.1	2	0	3	0	1
	1703.0	1703.1	2	0	2	2	1
	1853.1	1853.2	2	0	3	1	1
	1855.2	1855.2	2	0	2	3	1
	2003.2	2003.1	2	0	4	0	1
Trimers	1881.2	1881.3	3	0	2	0	2
	2335.5	2336.2	3	0	3	1	2
	2485.6	2485.3	3	0	4	0	2
	2487.6	2487.3	3	0	3	2	2
	2639.7	2639.3	3	0	3	3	2
	2787.8	2787.4	3	1	2	2	2
Tetramers	3270.1	3270.3	4	0	5	0	3
	3272.1	3272.3	4	0	4	2	3
Pentamers	4054.7	4054.8	5	0	6	0	4
	4056.7	4056.3	5	0	5	2	4

Data represents the observed and calculated masses of ellagitannin and other hydrolysable tannins in positive reflectron mode detected in pomegranate fruit extract as [M + Cs]$^+$ ions. Masses were calculated from the number of monomeric units shown in the table.

1.3.2.2 LC-ESI Mass Spectrometry

Coupling of high-performance liquid chromatography (HPLC) with mass spectrometry (MS) equipped with mild ionization techniques such as electrospray ionization (ESI) allows for the study of complex mixtures by the association of one separation method and one identification method. In ESI-MS the ionization process is done at room temperature and at atmospheric pressure. ESI-MS also allows the detection of multicharged ions in the positive ion mode $[M + zH]^{z+}/z$ or negative ion mode $[M - zH]^{z-}/z$. Besides, increasing the orifice voltage in the ion source favors the formation of fragment ions (MS^n), which gives supplementary information about the structure of the molecule. Fragment ions can be used to determine molecular structure as they arise from well-determined breaking of chemical bonds.

Tandem mass spectrometry (MS^n) and ion-trap mass spectrometry (TI-MS, MS^n) which allows fragmentation patterns to be obtained on selected individual ions, are progressively replacing the classical ESI-quadrupole mass spectrometers in HPLC-MS coupling. This facilitates the structural elucidation of unknown compounds. In addition, the utilization of the select ion monitoring mode (SIM) considerably increases the sensitivity. Recently, ellagitannins were detected as doubly charged ions by LC/ESI-MS in the negative ion mode.[48,49] The distinction between mono-charged and doubly charged signals depends on the distance between the isotope peaks due to the natural occurrence of carbon 13 (1%), as the spacing between the isotopic peaks is equal to one mass unit for monocharged ions and 0.5 mass units for doubly charged ions.[48,49] Mullen et al. produced a basis for the interpretation of fragmentation patterns (MS^n) from oligomeric ellagitannins.[49] Fragmentation across ester bonds gave rise to masses representing the loss of monomeric units: M-302, loss of a hexahydroxydiphenoyl (HHDP) group; M-162, loss of a glucosyl group; M-170, loss of a gallate group. In the case of pomegranate oligomeric HTs, fragmentation of punicalin and punicalagin would also be expected to give rise to masses representing gallagic acid and monomeric units. The LC-MS^n method is also useful in distinguishing between conjugates of quercetin and ellagic acid since their aglycons produce identical molecular ions on fragmentation ($M - H^+$, m/z 301).[48,49] In MS/MS analyses, in negative mode, the quercetin m/z 301 ion further fragments to form characteristic m/z 179 and 151 ions whereas the equivalent ellagic acid m/z 301 ion yields ions at m/z 257 and 229.[48,49] The use of LC-MS^n methods is therefore useful in differentiating between EA and quercetin aglycons, phytochemicals that are common to pomegranates.

1.3.2.3 Gas Chromatography-Mass Spectroscopy (GC-MS)

Gas chromatography with flame ionization and mass spectrometric detectors have been utilized for the identification of lipid-soluble pomegranate phytochemicals such as fatty acids, triglycerides, and other components of its seed and seed oil.[50] The use of GC and GC-MS usually requires derivatization of samples before they can be analyzed.

1.3.3 Nuclear Magnetic Resonance (NMR)

Liquid chromatographic separation techniques coupled with mass spectrometry, alone, are not adequate to elucidate structural confirmation. Mass spectra do not distinguish between isomers and give no information regarding regiochemistry of C–C and C–O bonds. Isolation of individual compounds and their isomers and characterization by ^{1}H and ^{13}C- NMR is necessary to determine absolute structure of pomegranate phytochemicals such as its HTs.[1,2,7]

1.3.4 Hydrolysis

Other chemical methods are also used for sample purification before complex pomegranate phytochemicals can be analyzed by sensitive mass spectral methods. For example, acidic and basic hydrolysis or enzymatic hydrolysis with tannase can be used to provide more detailed information on structural composition of pomegranate phytochemicals.[1,2] The hydrolysate can then be analyzed by HPLC, MALDI-TOF-MS, or LC-ESI-MS to determine the monomeric subunits of the oligomeric ellagitannins.

1.4 ROLE OF POMEGRANATE FRUIT PHYTOCHEMICALS IN BIOACTIVITY

The scope of this section is to review the current knowledge on the bioactivities of phytochemicals found in the pomegranate fruit. The emphasis is placed on those studies relating the observed bioactivities (in commercial or freshly prepared juice, peels, and seeds, with or without prefermentation) to specific phytochemical constituents of the pomegranate, even where such linkages have not yet been proven. The major *in vitro* and *in vivo* bioactivities evaluated to date have been the antioxidant and antiatherogenic properties of pomegranate juice (either fresh, commercial, or fermented), its extracts, and their purified compounds.

1.4.1 Pomegranate Fruit Phytochemicals As Antioxidants and Antiatherogenic Agents

1.4.1.1 *In Vitro* Studies

In vitro antioxidant activities in different parts of the pomegranate fruit, described in a number of reports and groundbreaking research initiated and conducted in the Lipid Research Laboratory of Professor Michael Aviram (Haifa, Israel), is discussed in a separate chapter in this book. Some other studies are described below.

Employing the fermentation process, PJ with seeds was first fermented for 10 days and then concentrated to one-tenth its original volume to yield pomegranate fermented juice (PFJ) extract.[50] The remaining seeds were separated, dried, and cold-pressed to form an additional product (PCPSO). The antioxidant activities of the PFJ and PCPSO were tested by the β-carotene linoleic acid method, as was their tendency to inhibit cyclooxygenase and lipoxygenase, key enzymes in the eicosanoid pathway catalyzing the oxidation of polyunsaturated fatty acids. The β-carotene linoleic acid method is based on determining the coupled oxidation of β-carotene and linoleic acid, then measuring β-carotene consumption as a result of its oxidation in the reaction mixture.[51,52] Both fermented products demonstrated moderate activity, less than the commercial synthetic antioxidant butylated hydroxyl anisole, and more than red wine. PFJ was not active as a cyclooxygenase inhibitor and only weakly inhibited lipoxygenase, while PCPSO revealed weak activity against cyclooxygenase and a strong inhibitory effect on lipoxygenase. As previously discussed, the major fatty-acid constituent in PCPSO is punicic acid, which is found in pomegranate seed oil (>60% of the total oil).

The antioxidant activity of polyphenols isolated from pomegranate seeds was determined by monitoring LDL susceptibility to oxidation and malondialdehyde (MDA) levels in the rat brain *in vitro*.[11] Two new compounds were isolated and identified as coniferyl and sinapyl glycoside derivatives. These two compounds moderately decreased conjugated diene (CD) formation and significantly inhibited MDA production in the rat brain in a dose-dependent manner. The other known polyphenols isolated from the seeds were ellagic acid and its derivatives, which also demonstrated appreciable antioxidant activity.

The antioxidant activity in PJ that was freshly prepared from arils was compared to that in commercial PJ extracted from whole pomegranate, red wine, and green tea infusion.[32] Antioxidant activity was determined using four different assays, three of them based on evaluations of the free-radical scavenging capacity of the juice (ABTS, DPPH, N,N-dimethyl-p-phenylenediamine-DMPD), and the fourth based on measuring its iron-reducing capacity (FRAP). The commercial juice revealed the highest activity, three times higher than that of red wine or green tea, and twice the antioxidant capacity of juice from fresh arils. An HPLC chromatogram of the two PJs showed that the commercial PJ includes phenolics additional to those present in the juice obtained from the arils, with a high content of punicalagins and ellagic acid derivatives. These results support the notion that the method of juice extraction plays an important role in its constitution and thus in its activity.[32] The commercial juice had a significantly higher total phenolics content (2560 mg/L) relative to juice from either frozen or fresh arils (1800 and 2120 mg/L, respectively), consisting mostly of punicalagins and ellagic acid derivatives.

The antioxidant activity as determined by the β-carotene, linoleic acid, and DPPH systems of extract from pomegranate peel was compared with that from seeds. The methanolic extract has been found to be superior to that obtained with ethyl acetate or water.[53] The methanolic peel extract was further tested as an inhibitor of lipid peroxidation, as a hydroxyl radical scavenger, and as an inhibitor of LDL oxidation. The antioxidant potency of the different peel extracts was correlated with their polyphenol content. However, no information was provided as to the type of constituents responsible for those activities, although the extracts were reported to contain ellagic and gallic acids that may probably be produced from the hydrolysis of ellagitannins and gallotannins, respectively.

1.4.1.2 *In Vivo* Studies

Among the *in vivo* experiments aimed at evaluating the biological effects of pomegranate juice (PJ), Aviram et al.[54] tested the effect of PJ consumption in male healthy and atherosclerotic, apolipoprotein E-deficient (E^0) mice. The authors demonstrated that PJ consumption has antiatherogenic properties with respect to all three related components of atherosclerosis: it significantly affected plasma lipoproteins, arterial macrophages, and blood platelets, all of which was attributed to the effects of the PJ's antioxidant constituents, specifically to a fraction containing ellagitannins.[54] In another study, Aviram and Dornfeld[55] tested the effect of PJ consumption by hypertensive patients on their blood pressure and on serum angiotensin-converting enzyme (ACE) activity. Hypertension is a known risk factor for the development of atherosclerosis, whereas inhibitors of ACE, an enzyme facilitating the conversion of angiotensin I to angiotensin II, have been shown to attenuate the development of atherosclerosis in several animal studies.[56] PJ consumption for 2 weeks caused a minimal reduction in blood pressure, but significantly decreased ACE activity by 36%; moreover, a small (5%) but significant reduction in systolic blood pressure was anticipated. It was assumed that the inhibition of ACE activity was either due to the direct action of specific inhibitors of ACE present in the juice and/or a secondary effect of PJ's antioxidants. In a separate randomized, double-blind, placebo-controlled study, PJ consumption for 3 months was shown to decrease stress-induced myocardial ischemia and improve myocardial perfusion in 45 patients with ischemic cardiovascular disease.[69]

Perturbed shear stress may trigger signal-transduction events that, in turn, can lead to endothelial dysfunction and enhanced atherogenesis.[57] De Nigris and colleagues tested, *in vitro* and *in vivo,* the effect of PJ on oxidation-sensitive genes and on endothelial nitrous oxide synthase (eNOS) expression, induced by high shear stress, using cultured human coronary artery endothelial cells (EC) and hypercholesterolemic mice.[58] Administration of PJ, diluted in the cellular medium of cultured human coronary artery EC exposed to laminar shear stress for 24 h, resulted in the reduced activation of redox-sensitive transcription factors (ELK-1, and *p*-JUN) and increased eNOS expression, both effects associated with antiatherogenic activity. Supplementation of PJ to hypercholesterolemic mice under oxidative stress (a high-fat diet) resulted in significantly lower plasma lipid peroxidation, a reduction in macrophage foam cell formation, and a 20% decrease in lesion area in atherosclerotic-prone regions, consistent with previous findings.[54] A second group of hypercholesterolemic mice, which was first allowed to develop the disease for 6 months and only then given PJ in their drinking water for 24 weeks, showed significantly reduced atherosclerotic progression, indicating that the proatherogenic effects induced by perturbed shear stress can be reversed by prolonged administration of PJ. These findings may have implications for the prevention or treatment of atherosclerosis. All of the aforementioned effects were attributed to polyphenol-rich PJ constituents, mainly its ellagitannins and anthocyanins.

Shear stress also mediated nuclear factor-kappa B (NF-κB) activation in vascular EC: such activation is associated with several pathologies, including atherosclerosis. Schubert et al. studied the ability of PJ fermented with wine yeast and then dealcoholized and concentrated (PW) to inhibit tumor necrosis factor α (TNF-α) and NF-κB activation in vascular EC.[59] PW proved to be a potent inhibitor of NF-κB activation; furthermore, it was shown that different antioxidants might have similar effects on TNF-α and NF-κB activation, but not necessarily through similar mechanisms of action.

Cerda et al. investigated the effects of supplementation with 1 L of juice/day (more than 5 g/day polyphenols, including 4.4 g/day punicalagin isomers) on rats by assaying a large number of hematological and serobiochemical parameters, such as LDL, high-density lipids (HDL), triglycerides and cholesterol, plasma and urine antioxidant activities (ability to donate electrons), and the presence of certain pomegranate constituents.[60] None of the major polyphenols in the original juice (ellagitannins, ellagic acid derivatives, and anthocyanins) were detected in the human plasma or urine. The most potent *in vitro* antioxidant in the original PJ, punicalagin, was not detected (intact) in the human plasma or urine, nor was punicalin or ellagic acid conjugates. It is noteworthy that in a separate study, intact ellagic acid, resulting from the *in vivo* hydrolysis of ellagitannins, was detected in human plasma after the consumption of PJ.[61] Cerda et al.[60] also reported that despite the high antioxidant activity of the original juice *in vitro,* supplementation of PJ to healthy human subjects did not show any parallel significant increase in antioxidant activity of the plasma or urine, as assayed by various parameters. These results are in contrast to previous data showing increased plasma antioxidant capacity after PJ supplementation.[54] The differences in results may be explained by the following observations: (a) the major known potent antioxidants, present in the original juice, were absent in the plasma or urine of the PJ-supplemented subjects, possibly due to their degradation and metabolism, (b) the antioxidant activity of the microfloral metabolites identified in the plasma or urine of those that consumed PJ are very weak, and (c) differences in antioxidant assay methodologies. For example, 3,8-dihydroxy-6H-dibenzo-pyran-6-one (urolithin A), an ellagitannin metabolite, showed 42- and 3570-fold lower activity than the parent compound, as assayed by the DPPH and ABTS antioxidant methods.[60] Other major metabolites were identified as glucuronide derivatives, with also undetectable antioxidant activity.[60]

Both Aviram et al. and Cerda et al. agreed that pomegranate supplementation decreases LDL cholesterol and β-lipoprotein levels, whereas other atherogenic factors, such as VLDL cholesterol and LDL triglycerides, increase. The latter report concluded that supplementation of 1 L of PJ for 5 days to healthy volunteers does not result in any conclusive evidence of incurred health benefits with regards to antioxidative capacity. These authors also called for caution concerning the output of data regarding the relevance of *in vitro* antioxidant capacity of foodstuffs to their *in vivo* extrapolation.[54,60]

The effect of feeding a methanolic extract of pomegranate peel to albino rats on the toxic effects of carbon tetrachloride (CCl_4) was examined using biochemical and histopathological assays.[62] Analysis of the methanolic extract showed 42% (w/w) of the total phenol as catechin equivalents, with gallic and ellagic acids as the two major constituents. The levels of various reactive-oxygen-species (ROS)-combatting enzymes, such as catalase, superoxide dismutase (SOD), and peroxidase, as well as the amount of lipid peroxide in the liver homogenates, were examined. Results were compared with the effect of CCl_4 on rats prefed the pomegranate peel extract. CCl_4 and its metabolites have been extensively studied as liver toxicants. CCl_4 reduced the levels of the aforementioned enzymes by 50 to 90%, and lipid peroxide increased about threefold. Pretreating the rats with the extract preserved the enzymes at control levels and reduced the lipid peroxide value to half of its control level. These protective effects could be attributed to the extract's high content of gallic and ellagic acids and possibly their polymers, all being potent free-radical scavengers.

In a recent study on arteriogenic erectile dysfunction done in a rabbit model, animals consuming PJ showed increased blood flow to the penis compared to control animals.[70] In addition, the time to maximum pressure (maximum erection) decreased compared to controls, and nitric-oxide (NO)-mediated smooth-muscle relaxation was improved in the treatment animals. The authors attributed the effects to the antioxidant phytochemicals present in pomegranates.

1.4.2 Neuroprotective Properties

There is growing evidence that pomegranate phytochemicals may play a role in neuroprotection in a variety of disease states. Recently, PJ has been shown to provide neuroprotection in an animal model of neonatal hypoxic-ischemic encephalopathy.[71] In this study, mice born to mothers who consumed PJ had less brain damage than controls and showed a reduction in caspase-3-activation. In addition, in this study, ellagic acid was found in the serum of animals afforded neuroprotection. Additional evidence of pomegranate's neuroprotective capabilities has been suggested by estrogen's attenuation of neonatal hypoxic-ischemia in an animal model.[72]

1.4.3 Anticarcinogenic Properties

The inhibitory activity of pomegranate seed oil (PGO) against colon cancer in rats was recently evaluated by Kohno et al.[63] Colon cancer is one of the leading causes of cancer deaths in Western countries. As previously discussed, pomegranate seed oil is rich in punicic acid, c9,t11,c13-conjugated linolenic acid (CLN), which makes up about 75% of the total linolenic acid in PGO. In an earlier study, Kohno et al. demonstrated that feeding rats CLN, isolated from bitter gourd, causes a significant reduction in colonic aberrant crypt foci (ACF) induced by azoxymethane.[64] These authors suggested that dietary intake of CLN may have an inhibitory effect on colon carcinogenesis. Their results indicated that dietary administration of PGO rich in c9,t11,c13-CLN, even at the low dose of 0.1% CLN, significantly inhibits the development of azoxymethane-induced colonic adenocarcinoma in rats without causing any adverse effects. Such results may reflect the potential chemopreventive effect of PGO on chemically induced colon carcinogenesis.

PJ, a purified pomegranate tannin extract, ellagic acid, and punicalagin have been shown to inhibit the proliferation of human oral, colon, and prostate tumor cells as well as show proapototic effects.[65] In this study, when the compounds were combined, enhanced effects were observed, which was attributed to additive and/or synergistic effects.

1.4.4 Phytoestrogenic Properties

Phytoestrogens are naturally occurring compounds present in various plants, such as soy and licorice,[66,67] that possess some of the activities of the female hormone

17-estradiol. The phytoestrogenic activity of pomegranate peel extract was determined by means of an online biochemical detection system combined with LC-MS.[17] This system merges the capacity of certain components in the extract under study to bind to β-estrogen receptor, while simultaneously separating and identifying the active components. Using the above method, three active flavonoids were found to be present in pomegranate peel extract and to bind to β-estrogen receptor-luteolin, quercetin, and kaempferol. It is interesting to note that when glycosylated derivatives of these three active phytoestrogens, which are also present in the extract, were examined separately, the estrogen-like activity demonstrated in the parent compounds was abolished. In another study, the well-known phytoestrogen coumestrol was detected at 0.036 mg/100g pomegranate seeds.[39]

1.4.5 POMEGRANATE SEED EXTRACT AS A TREATMENT FOR DIARRHEA

Methanolic extract from pomegranate seeds was found to reduce diarrhea by inhibiting gastrointestinal motility and PGE_2-induced enteropooling.[68] It was speculated that the phytochemicals responsible for this activity are the tannins present in the extract, which react with proteins to form tannate, causing denaturation of the original protein and thus reducing secretions from the intestinal mucosa.

1.4.6 METHANOLIC EXTRACT OF POMEGRANATE PEEL EXHIBITS WOUND-HEALING ACTIVITY

Methanolic extract of pomegranate peel, after fatty-acid removal (by preextraction with hexane), exhibited good healing abilities, with acceleration of wound progression.[62] Employing a Wistar rat skin model, 56% wound healing was achieved by day 15 following application of 2.5% of the extract in a water-soluble gel formulation, which is comparable to the commercial drug preparation (silver sulfadiazine). Assuming that the effect was due to the extract being rich in polyphenols, it was speculated that these polyphenols have the capacity to interact with proteins and to precipitate them from animal hide, thereby quickening the healing process.

1.5 SUMMARY AND CONCLUSIONS

It is paramount that researchers continue to develop novel methods of chromatographic separation to be used in conjunction with phytochemical analyses such as mass spectral analysis and techniques of NMR to improve our understanding of the composition of pomegranate phytochemicals, for among other things its oligomers. In addition to improving the detection capabilities of mass spectral analysis, the techniques of separation may be used by biomedical researchers to produce well-defined fractions of oligomers for use in *in-vivo* and *in-vitro* bioassays, improving their ability to relate structure to function.

Research aimed at investigating the beneficial effects of fresh, commercial, and fermented pomegranate fruit juice, peel, and seeds (and their purified single phytochemicals) is relatively new, despite the fact that the fruit has been consumed for thousands of years. Most of the *in vivo* studies have been aimed at evaluating the antioxidant and antiatherogenic properties of PJ, either directly (by measuring the size of the atherosclerotic lesions, or the extent of foam-cell formation), or indirectly (by measuring such parameters as the level of plasma lipoproteins, blood platelets, activities of various enzymes associated with oxidative stress, and gene expression). However, these studies provide little information about the phytochemical constituents responsible for the observed activities, or about the bioavailability of the suspected active agents. The instances in which specific components of PJ were tested involved *in vitro* experiments, and as such their contribution to an exploration of the role of specific molecules in pomegranate fruits is less significant. The tendency for this type of research (i.e., in which the *in vivo* effect of the whole fruit is investigated) to become dominant may have several explanations: it is a relatively young field, giving it a higher chance of yielding positive results and a lower chance of running into legislative difficulties. In the coming years, however, we may see the gap narrow, with more effort being directed at investigating the effects of specific pomegranate phytochemicals and their biological fate *in vivo,* as well as research directed at exploring other biological effects of these phytochemicals and their mechanisms of action.

REFERENCES

1. Tanaka, T., Nonaka, G., and Nishioka, I., Tannins and related compounds. XL. Revision of the structures of punicalin and punicalagin, and isolation and characterization of 2-O-galloylpunicalin from the bark of *Punica granatum* L., *Chem. Pharm. Bull.,* 34, 650, 1986.
2. Tanaka, T., Nonaka, G., and Nishioka, I., Tannins and related compounds. XLI. Isolation and characterization of novel ellagitannins, punicacorteins A, B, C and D, and punigluconin from the bark of *Punica granatum* L, *Chem. Pharm. Bull.,* 34, 656, 1986.
3. Satomi, H. et al., Carbonic anhydrase inhibitors from the pericarps of *Punica granatum* L, *Biol. Pharm. Bull.,* 16, 787, 1993.
4. Mavlyanov, S.M. et al., Polyphenols of the fruits of some varieties of pomegranate growing in Uzbekistan, *Chem. Nat. Compd,* 33, 98, 1997.
5. El-Toumy, S.A.A., Marzouk, M.S., and Rauwald, H.W., Ellagi- and gallotannins from *Punica granatum* heartwood, *Pharmazie,* 56, 823, 2001.
6. Haddock, E.A., Gupta, R.K., and Haslam, E., The metabolism of gallic acid and hexahydroxydiphenic acid in plants. Part 3. Esters of (R) and (S)-hexahydroxydiphenic acid and dehydrohexahydroxydiphenic acid with D-glucopyranose (1C4 and related conformations), *J. Chem. Soc., Perkin Trans. 1,* 2535, 1982.
7. Tanaka, T., Nonaka, G., and Nishioka, I., Tannins and related compounds. Part 30. Punicafolin, an ellagitannin from the leaves of *Punica granatum, Phytochemistry,* 24, 2075, 1985.
8. Hussein, S.A.M. et al., Tannins from the leaves of *Punica granatum, Phytochemistry,* 45, 819, 1997.
9. El-Toumy, S.A.A. and Rauwald, H.W., Two ellagitannins from *Punica granatum* heartwood, *Phytochemistry,* 61, 971, 2002.
10. Nawwar, M.A.M. and Hussein, S.A.M., NMR spectra analysis of polyphenols from *Punica granatum, Phytochemistry,* 36, 793, 1994.

11. Wang, R. et al., Bioactive compounds from the seeds of *Punica granatum* (Pomegranate), *J. Nat. Prod.,* 67, 2096, 2004.
12. Artik, N. et al., Determination of phenolic compounds in pomegranate juice by HPLC, *Fruit Processing,* 8, 492, 1998.
13. Plumb, G.W. et al., Antioxidant properties of gallocatechin and prodelphinidins from pomegranate peel, *Redox Rep.,* 7, 41, 2002.
14. Hernandez, F. et al., Evolution of juice anthocyanins during ripening of new selected pomegranate *(Punica granatum)* clones, *Eur. Food Res. and Tech.,* 210, 39, 1999.
15. Santagati, N.A., Duro, R., and Duro, F., Study on pigments present in pomegranate seeds, *J. Commodity Sci.,* 23, 247, 1984.
16. Nawwar, M.A.M., Hussein, S.A.M., and Merfort, I., Leaf phenolics of *Punica granatum, Phytochemistry,* 37, 1175, 1994.
17. van Elswijk, D.A. et al., Rapid dereplication of estrogenic compounds in pomegranate *(Punica granatum)* using on-line biochemical detection coupled to mass spectrometry, *Phytochemistry,* 65, 233, 2004.
18. Lugasi, A. and Hovari, J., Flavonoid aglycons in foods of plant origin II. Fresh and dried fruits, *Acta Aliment. Hung,* 31, 63, 2002.
19. Chauhan, D. and Chauhan, J.S., Flavonoid diglycoside from *Punica granatum, Pharm Biol.,* 39, 155, 2001.
20. Srivastava, R., Chauhan, D., and Chauhan, J.S., Flavonoid diglycosides from *Punica granatum, Indian J. Chem., Section B,* 40B, 170, 2001.
21. Poyrazoglu, E., Goekmen, V., and Artik, N., Organic acids and phenolic compounds in pomegranates *(Punica granatum* L.) grown in Turkey, *J. Food Composition and Anal.,* 15, 567, 2002.
22. Chen, Y. et al., Study on fatty acids extracted from pomegranate seed oil, *Shipin Kexue,* 24, 111, 2003.
23. Yusuph, M. and Mann, J., A triglyceride from *Punica granatum, Phytochemistry,* 44, 1391, 1997.
24. Fatope, M.O., Al Burtomani, S.K.S., and Takeda, Y., Monoacylglycerol from Punica granatum seed oil, *J. Agric Food Chem.,* 50, 357, 2002.
25. Batta, A.K. and Rangaswami, S., Crystalline chemical components of some vegetable drugs, *Phytochemistry,* 12, 214, 1973.
26. Brieskorn, C.H. and Keskin, M., Betulic acid in the leaves of *Punica granatum, Pharm. Acta Helv.,* 30, 361, 1955.
27. El Wahab, S.M. et al., Characterization of certain steroid hormones in *Punica granatum* L. seeds, *Bulletin of the Faculty of Pharmacy,* 36, 11, 1998.
28. Fayez, M.B.E., Negm, S.A.R., and Sharaf, A., Constituents of local plants. V. The constituents of various parts of the pomegranate plant, *Planta Med.,* 11, 439, 1963.
29. Ahmed, R. et al., Studies on *Punica granatum.* I. Isolation and identification of some constituents from the seeds of *Punica granatum, Pakistan J. Pharm. Sci.,* 8, 69, 1995.
30. Neuhoefer, H. et al., Alkaloids in the bark of *Punica granatum* L. (pomegranate) from Yugoslavia, *Pharmazie,* 48, 389, 1993.
31. Fayez, M.B.E., Negm, S.A.R., and Sharaf, A., Constituents of local plants. V. The constituents of various parts of the pomegranate plant, *Planta Med.,* 11, 439, 1963.
32. Gil, M.I. et al., Antioxidant activity of pomegranate juice and its relationship with phenolic composition and processing, *J. Agric. Food Chem.,* 48, 4581, 2000.
33. Seeram, N.P. et al., Rapid large-scale purification of ellagitannins from pomegranate husk, a by-product of the commercial juice industry. *Sep. Purif. Technol.,* 41, 49, 2005.
34. Narr, B.C, Ayed, N., and Metche, M., Quantitative determination of the polyphenolic content of pomegranate peel. *Z Lebensm Unters Forsch,* 203, 374, 1996.

35. Haslam, E., *Plant polyphenols: Vegetable tannins revisited.* Cambridge University Press, Cambridge, 1989, 230.

36. Kohno, H. et al., Pomegranate seed oil rich in conjugated linolenic acid suppresses chemically induced colon carcinogenesis in rats. *Cancer Sci.,* 95, 481, 2004.

37. Melgarejo, P. et al., Total lipids content and fatty acid composition of seed oils from six pomegranate cultivars. *J. Sci. Food and Agric.,* 69, 253, 1995.

38. Heftmann, E., Ko, S.T., and Bennett, R.D., Identification of estrone in pomegranate seeds, *Phytochemistry,* 5, 1337, 1966.

39. Moneam, N.M., El Sharaky, A.S., and Badreldin, M.M., Oestrogen content of pomegranate seeds, *J. Chromatogr.,* 438, 438, 1988.

40. Choi, D.W. et al., Identification of steroid hormones in pomegranate *(Punica granatum)* using HPLC and GC-mass spectrometry, *Food Chem.,* (in press), 2005.

41. Krueger, C.G. et al., Matrix-assisted laser desorption/ionization time-of-flight mass spectrometry of polygalloyl polyflavan-3-ols in grape seed extract, *J. Agric. Food Chem.,* 48, 1663, 2000.

42. Krueger, C.G., Vestling, M.M., and Reed, J.D., Matrix-assisted laser desorption/ionization time-of-flight mass spectrometry of heteropolyflavan-3-ols and glucosylated heteropolyflavans in sorghum [*Sorghum bicolor* (L.) Moench], *J. Agric. Food Chem.,* 53, 538, 2003.

43. Krueger, C.G., Vestling, M.M., and Reed, J.D., Matrix-assisted laser desorption/ionization time-of-flight mass spectrometry of anthocyanin-polyflavan-3-ol oligomers in cranberry fruit (*Vaccinium macrocarpon,* Ait.) and spray dried cranberry juice, in *Red wine color: revealing the mysteries, ACS Symposium Series,* eds. A.L. Waterhouse and J.A. Kennedy, vol. 886, American Chemical Society, Washington, D.C., 2004, 232.

44. Reed, J.D., Krueger, C.G., and Vestling, M.M., MALDI-TOF mass spectrometry of oligomeric food polyphenols, *Phytochemistry,* 66, 2248, 2005.

45. Porter, M.L. et al., Cranberry proanthocyanidins associate with low-density lipoprotein and inhibit in vitro Cu^{2+}-induced oxidation, *J. Sci. Food and Agric.,* 81, 1306, 2001.

46. Afaq, F. et al., Anthocyanin and hydrolyzable tannin-rich pomegranate fruit extract modulates MAPK and NF-κB Pathways and inhibits skin tumorigenesis in CD-1 mice, *Internl. J. Cancer,* 113, 423, 2005.

47. Quideau, S. and Feldman, K., Ellagitannin chemistry, *Chem. Rev.,* 96, 475, 1996.

48. Seeram, N.P. et al., Identification of phenolics in strawberries by liquid chromatography electrospray ionization mass spectroscopy, *Food Chem.,* 2005 (in press).

49. Mullen, W. et al., Analysis of ellagitannins and conjugates of ellagic acid and quercetin in raspberry fruits by LC-MSn, *Phytochemistry,* 64, 617, 2003.

50. Schubert, S.Y., Lansky, E.P., and Neeman, I., Antioxidant and eicosanoid enzyme inhibition properties of pomegranate seed oil and fermented juice flavonoids, *J. Ethnopharmacol.,* 66, 11, 1999.

51. Aviram, M. and Vaya, J., Markers for low-density lipoprotein oxidation, *Meth. Enzymol.,* 335, 244, 2001.

52. Taga, M.S., Miller, E.E., and Pratt, D.E., Chia seeds as a source of natural lipid antioxidants, *J. Am. Oil Chem. Soc.,* 61, 928, 1984.

53. Murthy, K.N.C., Jayaprakasha, G.K., and Singh, R.P., Studies on antioxidant activity of pomegranate *(Punica granatum)* peel extract using *in vivo* models, *J. Agric. Food Chem.,* 50, 4791, 2002.

54. Aviram, M. et al., Pomegranate juice consumption reduces oxidative stress, atherogenic modifications to LDL, and platelet aggregation: studies in humans and in atherosclerotic apolipoprotein E-deficient mice, *Am. J. Clin. Nutr.,* 71, 1062, 2000.

55. Aviram, M. and Dornfeld, L., Pomegranate juice consumption inhibits serum angiotensin converting enzyme activity and reduces systolic blood pressure, *Atherosclerosis,* 158, 195, 2001.

56. Kowala, M.C., Grove, R.I., and Aberg, G., Inhibitors of angiotensin converting enzyme decrease early atherosclerosis in hyperlipidemic hamsters. Fosinopril reduces plasma cholesterol and captopril inhibits macrophage-foam cell accumulation independently of blood pressure and plasma lipids, *Atherosclerosis,* 108, 61, 1994.

57. Gimbrone, M.A., Jr., Vascular endothelium, hemodynamic forces, and atherogenesis, *Am. J. Pathol.,* 155, 1, 1999.

58. de Nigris, F. et al., Beneficial effects of pomegranate juice on oxidation-sensitive genes and endothelial nitric oxide synthase activity at sites of perturbed shear stress, *Proc. Natl. Acad. Sci. USA,* 102, 4896, 2005.

59. Schubert, S.Y., Neeman, I., Resnick, N., A novel mechanism for the inhibition of NF-kappaB activation in vascular endothelial cells by natural antioxidants, *Faseb J.,* 16, 1931, 2002.

60. Cerda, B. et al., The potent *in vitro* antioxidant ellagitannins from pomegranate juice are metabolised into bioavailable but poor antioxidant hydroxy-6H-dibenzopyran-6-one derivatives by the colonic microflora of healthy humans, *Eur. J. Nutr.,* 43, 205, 2004.

61. Seeram, N.P., Lee, R., and Heber, D., Bioavailability of ellagic acid in human plasma after consumption of ellagitannins from pomegranate (*Punica granatum* L.) juice, *Clin. Chim. Acta,* 348, 63, 2004.

62. Murthy, K.N. et al., Study on wound healing activity of *Punica granatum* peel. *J. Med. Food,* 7, 256, 2004.

63. Kohno, H. et al., Pomegranate seed oil rich in conjugated linolenic acid suppresses chemically induced colon carcinogenesis in rats, *Cancer Sci.,* 95, 481, 2004.

64. Kohno, H. et al., Dietary conjugated linolenic acid inhibits azoxymethane-induced colonic aberrant crypt foci in rats, *Jpn J. Cancer Res.,* 93, 133, 2002.

65. Seeram, N.P. et al., In vitro antiproliferative, apoptotic and antioxidant activities of punicalagin, ellagic acid and a total pomegranate tannin extract are enhanced in combination with other polyphenols as found in pomegranate juice, *J. Nutr. Biochem.,* 6, 360, 2005.

66. Cassidy, A., Bingham, S., and Setchell, K.D., Biological effects of a diet of soy protein rich in isoflavones on the menstrual cycle of premenopausal women, *Am. J. Clin. Nutr.,* 60, 333, 1994.

67. Tamir, S. et al., Estrogenic and antiproliferative properties of glabridin from licorice in human breast cancer cells, *Cancer Res.,* 60, 5704, 2000.

68. Das, A.K. et al., Studies on antidiarrhoeal activity of *Punica granatum* seed extract in rats, *J. Ethnopharmacol.,* 68, 205, 1999.

69. Sumner, M.D. et. al., Effects of pomegranate juice consumption on myocardial perfusion in patients with coronary heart disease. *Am. J. Cardiol.,* 96, 810, 2005.

70. Azadzoi, K.M. et al.. Oxidative stress in arteriogenic erectile dysfunction: prophylactic role of antioxidants, *J. Urology,* 174, 386, 2005.

71. Loren, D.J. et al., Maternal dietary supplementation with pomegranate juice is neuroprotective in an animal model of neonatal hypoxic-ischemic brain injury, *Pediatr. Res.* 57, 858, 2005.

72. Feng, Y., Fratkins, J.D., and LeBlanc, M.H., Estrogen attenuates hypoxic-ischemic brain injury in neonatal rats, *Eur. J. Pharmacol.,* 507, 77, 2005.

2 Antioxidative Properties of Pomegranate: *In Vitro* Studies

Mira Rosenblat and Michael Aviram

CONTENTS

2.1 NUTRITIONAL ANTIOXIDANTS AND POLYPHENOLIC FLAVONOIDS

The beneficial health effects attributed to the consumption of fruits and vegetables are related, at least in part, to their antioxidant activity.[1,2] Edible plants contain several hundred different antioxidants.[3] Natural antioxidants include vitamins C and E, carotenoids (such as β-carotene and tomato's lycopene), and polyphenolic flavonoids (such as those present in grapes, berries, licorice, ginger, nuts, and olive oil).

For a compound to be defined as an antioxidant it must satisfy two basic conditions:

1. When present at low concentration relative to the substrate to be oxidized, it can delay, retard, or prevent autooxidation or free radical-mediated oxidation.
2. The resulting radical formed after scavenging must be stable in order to interrupt the oxidation chain reaction.

Low-density lipoprotein (LDL) oxidation is considered to be a hallmark of early atherogenesis. Nutritional antioxidants can affect LDL oxidation directly or indirectly, via modulations of arterial wall cell oxidative state and its subsequent capacity to oxidize LDL. Dietary antioxidants can inhibit LDL oxidation by several means:

1. Scavenging of free radicals, chelation of transition metal ions, or protection of the intrinsic antioxidants in the LDL particle (vitamin E and carotenoids) from oxidation.
2. Protecting cells in the arterial wall against oxidative damage and, as a result, inhibition of cell-mediated oxidation of LDL.
3. Preservation of serum paraoxonase (a high-density lipoprotein [HDL]-associated enzyme, which hydrolyzes specific lipid peroxides) activity.

Vitamin E (α-Tocopherol) has been proposed to be a very important lipid-soluble radical-scavenging antioxidant in cellular and subcellular membranes and also in plasma lipoproteins.[4] *In vitro* studies demonstrated that vitamin E inhibits peroxyl-radical- or heme-induced LDL oxidation.[5,6] Vitamin E supplementation to apolipoprotein E-deficient (E^0) mice or to Watanabe Heritable Hyperlipidemic (WHHL) rabbits resulted in prolongation of the LDL oxidation lag time and attenuated atherosclerosis development.[7,8] In healthy humans, vitamin E supplementation for 2 months resulted in LDL resistance to oxidation[9], and in patients with carotid artery stenosis, short-term alpha-tocopherol supplementation (500 IU/day) also inhibited lipid peroxidation in the patients' atherosclerotic lesions.[10]

Carotenoids are natural pigments with lipophylic properties that also possess antioxidant characteristics.[11,12,13] β-carotene and lycopene are the major carotenoids in human plasma, and LDL supplementation with β-carotene or with lycopene increases its resistance to oxidation.[14] Lycopene acts as an effective antioxidant against LDL oxidation in synergy with several natural antioxidants, including vitamin E, the isoflavan glabridin, and the phenolics rosmarinic and carnosic acid.[15] Furthermore, enrichment of LDL *in vivo* with a mixture of vitamin E, β-carotene, lycopene, asthaxanthin, and lutein following a single oral supplementation resulted in the protection of LDL polyunsaturated fatty acids (PUFA) and cholesterol moieties against oxidative modification.[16] The consumption by healthy adults of food products containing moderate amounts of vitamin E and carotenoids lead to a significant reduction in lipid peroxidation.[17,18]

Flavonoids comprise the largest and most-studied group of plant antioxidants, with more than 4000 different flavonoids being identified to date. They are usually found in plants as glycosides, and large compositional differences exist between different types of plants, even between different parts of the same plant.[19] Flavonoids are grouped into anthocyanins and anthoxantins. Anthocyanins are glycosides of anthocyanidin and are the most important group of water-soluble plant pigments, responsible for the red, blue, and purple colors of flowers and fruits. Anthoxantins are colorless or colored white-to-yellow, and include flavonols, flavanols, flavones, flavans, and isoflavones.

Flavonoids are powerful antioxidants against LDL oxidation, and their activity is related to their localization in the LDL particle, as well as to their chemical structure.[20–24] Flavonoids are effective scavengers of hydroxyl and peroxyl radicals, as well as of superoxide anion.[25,26] Some act also as antioxidants due to their potent chelation capacity of transition metal ions.[26,27] Furthermore, flavonoids can preserve serum paraoxonase 1 (PON1) activity probably as a result of the reduction in the level of oxidized lipids.[28,29] Flavonoids are also quite suitable for protecting cell membranes from free-radical-induced oxidation, since they are both lipophilic and hydrophilic and thus are able to scavenge free radicals, which are generated within the cells, as well as free radicals that attack the cell membrane from the outside. Indeed, flavonoids were shown to reduce macrophage oxidative stress and cell-mediated oxidation of LDL.[30]

We have performed *in vitro* and *ex vivo* studies in humans and in the atherosclerotic E^0 mouse, and in a cell-free system, which demonstrated that the licorice root extract and its purified isoflavane glabridin protected LDL against macrophage-induced oxidation.[31–35] Flavonols and their glycosides are effective antioxidants against 2,2′-azobis 2-amidino-propane dihydrochloride (AAPH- free-radical generator) and copper ion-induced LDL oxidation. The flavonols bearing the orthodihydroxyl groups possess significantly higher antioxidative activity than those bearing no such functionalities. The flavonol glycosides are less effective in this respect.[36]

Red wine and its flavonol quercetin were shown to inhibit LDL oxidation in both water and lipid-soluble free-radical-generating systems.[37,38] We have previously shown that preparation of white wine from whole squeezed grapes in the presence of 18% alcohol remarkably increased the extraction of grape skin polyphenols into the produced white wine, paralleled by an improvement of the white wine antioxidant activity, up to a similar extent to that present in red wine.[39] Red wine consumption by healthy subjects increased the resistance of LDL to oxidation,[40,41] and this effect depends on the polyphenol composition of the red wine.[42] Grape seeds extract also possesses antioxidative properties, as it was shown to reduce LDL oxidation in heavy smokers, who are under oxidative stress.[43] In E^0 mice under oxidative stress, supplementation of grape powder polyphenols, or red wine or its polyphenols quercetin or catechin, decreased LDL oxidation and attenuated atherosclerosis development.[44–47]

The pomegranate tree, which is said to have flourished in the Garden of Eden, has been extensively used as a folk medicine in many cultures. The fruit of the pomegranate (~50% of total pomegranate weight) consists of 80% juice and 20% seeds. The fresh juice contains 85% water, 10% total sugars, 1.5% pectin, ascorbic acid, and polyphenolic flavonoids.[48,49] In pomegranate juice (PJ), fructose and glucose are present in similar quantities (but it contains additional complexed, as-yet-unidentified sugars), calcium is 50% of its ash content, and the principal amino acids are glutamic and aspartic acids.[50] The soluble polyphenol content varies within the limits of 0.2 to 1.0%, and includes mainly ellagic tannins, gallic and ellagic acids, anthocyanins, and catechins.[51,52] Consumption of PJ by humans (one 8-oz [240-ml] glass a day) for a period of one year significantly increased PON1 activity and reduced the oxidation of both LDL and HDL by up to 60%. Most of these effects were already achieved in the first month of juice consumption.[53] Studies of patients with carotid artery stenosis (blockage in the arteries that supply blood to the brain) who consumed PJ for 3 years clearly demonstrated reduced oxidative stress in their blood, increased PON1 activity (by 83%), and, most importantly, a decreased atherosclerotic lesion size (by 30%).[54] Similarly, in E^0 mice supplemented with PJ, serum oxidative stress, macrophage-mediated oxidation of LDL, and the progression of atherosclerotic lesion were significantly inhibited.[53,55] Similarly, in a recent study, oral administration of PJ to hypercholesterolemic mice at various stages of the disease significantly reduced the progression of atherosclerosis. This effect of PJ could be attributed to its effect on oxidation-sensitive genes and on endothelial NO synthase activity at the site of perturbed shear stress.[56]

2.2 ANTIOXIDATIVE PROPERTIES OF THE POMEGRANATE TREE PARTS

Since pomegranate tree parts may also contain polyphenols with antioxidant activity,[57] we have prepared ethanolic extracts of whole crude pomegranate plant parts. The polyphenol concentrations in these extracts decreased in the following order: bark > stem > whole fruit juice > leaves. In the 1,1-diphenyl-2-picrylhydrazyl (DPPH) free radical scavenging capacity assay, using similar total polyphenol content (10 μmol of total polyphenols/L) from each of the pomegranate tree parts, the following order of potency was observed: bark > whole fruit juice > stem > leaves. Analysis of the antioxidant activity against copper ion-induced LDL oxidation, using similar total polyphenol concentrations, revealed a dose-dependent inhibition of LDL oxidation. These results suggest that different polyphenolic flavonoids are present in the pomegranate tree parts and, hence, on using total polyphenol content, specific flavonoids exert different antioxidative potency. The pomegranate flower powder ethanolic extract also contained polyphenols (45 nmoles of total polyphenols/mg flowers weight), and was also able to scavenge free radicals and to substantially inhibit copper ion-induced LDL oxidation in a dose-dependent manner. In order to analyze the components in the different tree parts with the antioxidative properties, we isolated several polyphenolic fractions and found that tannins were the most powerful antioxidants in the various pomegranate tree parts (Figure 2.1). The tannin fractions from pomegranate bark and stem contained the highest polyphenol content (Figure 2.1a) and in parallel, demonstrated the highest antioxidative capacities against LDL oxidation (Figure 2.1b).

2.3 ANTIOXIDATIVE PROPERTIES OF THE POMEGRANATE FRUIT FRACTIONS

The antioxidant activities of the whole pomegranate fruit and its anthocyanidins (delphinidin, cyanidin, and pelargonidin) extracts were studied.[58] Pomegranate fruit extract exhibited scavenging activity against hydroxyl radicals and superoxide anions, and this effect could be related to the fruit anthocyanidins. The IC_{50} (the concentration needed to get 50% inhibition) values for superoxide anion scavenging capacity of delphinidin, cyanidin, and pelargonidin were 2, 22, and 456 μmol/L, respectively. The above anthocyanidins did not effectively scavenge NO radical, but inhibited H_2O_2-induced lipid peroxidation in rat brain homogenates, with IC_{50} values of 0.7, 3.5, and 85 μmol/L for delphinidin, cyanidin, and pelargonidin, respectively.[58]

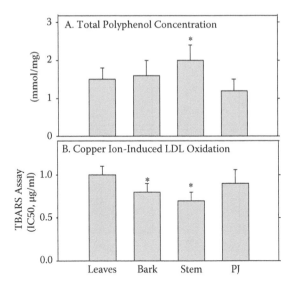

FIGURE 2.1 The tannin fractions were isolated from the following pomegranate *tree parts:* leaves, bark, stem, and whole-fruit pomegranate juice (PJ). The powder extracts were dissolved in ethanol. Total polyphenol concentration in the tannin fractions (a) was determined. LDL (100 µg of protein/mL) was incubated for 2 hours at 37°, with increasing weight concentrations of the different fractions in the presence of 5 µmol/L $CuSO_4$. The extent of LDL oxidation was measured by the TBARS assay. The IC_{50} (the concentration needed for 50% inhibition in LDL oxidation) is given (b). Results represent mean ± SD of three different experiments, *$p < 0.01$ the bark and the stem fractions vs. PJ fractions.

We compared the various pomegranate fruit constituents: aril juice, skin (red), membrane (white), and seeds. All these parts of the fruit contain polyphenols, with the highest concentration observed in the membrane fraction and the lowest concentration in the seed aqueous extract. Per 1 mg of weight the membrane contains 20- to 30-fold more polyphenols than the aqueous fractions of the seeds and aril juice. Based on an equal polyphenol concentration, we demonstrated that the aqueous extract of the pomegranate membranes was a more powerful antioxidant, in comparison to the aril juice, suggesting that they contain more potent polyphenol antioxidants. The concentrations of polyphenols, which were required to inhibit LDL oxidation by 50% (IC_{50}), were 0.6 µmol/L for the membrane compared to 1.0 µmol/L for the aril juice. Similar results were obtained upon comparing pomegranate peel and juice extracts to other foodstuffs.[2] In this study the order of antioxidant activity was pomegranate peel > turmeric > ragi > amla > pomegranate juice > amaranth > rajmah > sesame > wheat > flaxseed. The aqueous extract obtained from the crushed seeds was found to be a weak antioxidant against LDL oxidation. Similar results were noted upon comparing methanol extracts of the pomegranate peels vs. seeds.[59] Cold-pressed seed oil showed antioxidant activity close to that of butylated hydroxyanisole and green tea, and significantly greater than that of red wine.[60]

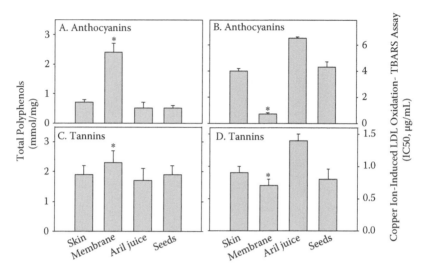

FIGURE 2.2 The anthocyanin and tannin fractions were isolated from the following pome-granate *fruit parts:* skin, membrane, aril juice, and seeds. The powder extracts were dissolved in ethanol. Total polyphenol concentrations in the anthocyanin fractions (a) and in the tannin fractions (c) were determined. LDL (100 g of protein/mL) was incubated for 2 hours at 37°, with increasing weight concentrations of the different fractions in the presence of 5 µmol/L $CuSO_4$. The extent of LDL oxidation was measured by the TBARS assay. The IC_{50} (the concentration needed for 50% inhibition in LDL oxidation) are given for the anthocyanin fractions (b) and for the tannin fractions (d). Results represent mean ± SD of three different experiments, $*p < 0.01$ vs. aril juice.

Anthocyanin and tannin fractions were next isolated from the skin, membrane, aril juice, and seeds. The fruit membrane anthocyanin and tannin fractions contained the highest polyphenol concentration (Figures 2.2a and 2.2c) and were the most potent antioxidants, with the lowest weight concentrations required for the inhibition of copper ion-induced LDL by 50% (IC_{50}, Figures 2.2b and 2.2d). Flavonoids extracted from the seed oil were shown to inhibit the activity of cyclooxygenase and of lipoxygenase (cellular oxygenases that affect oxidative stress).[60] Derivatives of glucopyranosides and of methylellagic acid that were isolated from pomegranate seeds significantly reduced LDL susceptibility to oxidation,[61] and prodelphinidin-peel-derived dimers were found to be much more effective antioxidants than another pomegranate peel anthocyanin, the gallocatechin monomer.[62]

2.4 ANTIOXIDATIVE CAPACITY OF POMEGRANATE JUICE IN COMPARISON TO OTHER JUICES

Pomegranate juice was shown to possess an antioxidant activity that was three times higher than the antioxidant activity of green tea.[63] The antioxidant activity was higher in juice extracted from whole pomegranate than that of juice obtained from arils only, suggesting that the processing extracts some of the hydrolyzable tannins present in the fruit rind into the juice.

We have demonstrated that pomegranate juice contains a higher concentration of total polyphenols (5 mmol/L) in comparison to other fruit juices (orange, grapefruit, grape, cranberry, pear, pineapple, apple, and peach juices, which contain only 1.3 to 4 mmol/L of total polyphenols, Table 2.1). A similar pattern was noted for IC_{50} values, obtained for the inhibition of copper ion-induced LDL oxidation. While PJ exhibited a very low IC_{50} (0.06 µl/mL), the IC_{50} values for the other juices were in the range of 0.11 to 7.50 µl/mL (Table 2.1). The most potent antioxidant activity of PJ could be related to its high polyphenolic flavonoids content, as well as to the specific type of potent polyphenols present in PJ (specific hydrolyzable tannins).

Incubation of LDL (100 µg of protein/mL) with increasing concentrations of PJ (0 to 3.5 µmol polyphenols/L) resulted in a PJ dose-dependent inhibition of LDL oxidation induced by 5 µmol/L $CuSO_4$, or by AAPH (reduced levels of lipid peroxides and aldehydes, and prolongation of the lag time required for the initiation of LDL oxidation). Furthermore, PJ also significantly inhibited cell-mediated oxidation of LDL by the J774 A.1 macrophage cell line by up to 67% on using 3.5 µmol of PJ total polyphenols. We have demonstrated *in vitro* that PJ not only inhibited LDL oxidation, but also was able to reduce macrophage oxidative stress. We did this by adding PJ (75 µmol of total polyphenols) to J774A.1 macrophages for 4 hours,

---

TABLE 2.1
Comparison between Pomegranate Juice and Other Fruit Juices: Polyphenol Concentration and Antioxidant Activity

Free-Radical Scavenging Capacity (% OD Reduction)	Copper Ion-Induced LDL Oxidation (IC_{50}, µL/mL)	Total Polyphenols (mmol/L)	Concentrated Juice
95	0.06	5.0	Pomegranate
80	0.11	4.5	Red plum
47	0.70	3.3	Grape
47	1.00	2.5	Cranberry
70	0.33	2.2	Kiwi
11	1.60	1.6	Orange
16	1.40	1.5	Grapefruit
55	1.20	1.4	Apple
27	1.00	1.1	Pineapple
5	7.50	1.1	Pear
30	2.25	1.0	Peach

Total polyphenol concentration in the different juices was determined using quercetin as a standard. LDL (100 µg of protein/mL) was preincubated with increasing volume concentrations (0 to 25 µL) of the juices. Then, 5 µmol/L of $CuSO_4$ was added, and the LDL was further incubated for 2 hours at 37°C. The extent of LDL oxidation was measured by the thiobarbituric acid reactive substances (TBARS) assay, and the IC_{50} values (the concentration needed to get 50% inhibition) are given. Free-radical scavenging capacity was analyzed by the DPPH assay, and is given as % of absorbance reduced by 1 µL/mL of juice after 5 minutes of incubation.

which significantly reduced cellular oxidative stress, measured by the DCFH assay, by 50%. Recently, pomegranate wine was shown to also significantly reduce oxidative stress induced by TNF-α in endothelial cells, as well as the activation of NF-κB.[64]

2.5 ANTIOXIDATIVE POTENCY OF FRACTIONS ISOLATED FROM POMEGRANATE JUICE

Several fractions were isolated from pomegranate juice, including gallic acid, ellagic acid, tannins, total PJ anthocyanins, and specific anthocyanins such as cyanidin-3-0-β-glucopyranoside, cyanidin-3,5-di-0-β-glucopyranoside, delphinidin-3-0-β-glucopyranoside, and pelargonidin-3-0-β-glucopyranoside. The total anthocyanin and tannin fractions exhibited a dose-dependent antioxidative effect against copper ion-induced LDL oxidation. In the AAPH-induced LDL oxidation, both fractions exhibited weaker antioxidative properties in comparison to the copper ion-induced LDL oxidation. These results suggest that the anthocyanins and tannins also possess transition metal ion chelation properties in addition to their free-radical scavenging capabilities. The tannin fraction was more potent than the anthocyanin fraction in inhibiting LDL oxidation, and the IC_{50} of the tannins was half that of the anthocyanins. Both PJ ellagic and gallic acids and the anthocyanins delphinidin-3-0-β-glucopyranoside, pelargonidin-3-0-β-glucopyranoside, cyanidin-3-0-β-glucopyranodise, and cyanidin-3,5-di-0-β-glucopyranoside inhibited copper ion-induced LDL oxidation in a dose-dependent manner. Upon comparing the effects of ellagic acid to gallic acid, gallic acid was more potent (IC_{50} of 2.1 μg/mL for gallic acid vs. 16 μg/mL for ellagic acid). Similarly, the anthocyanins delphinidin-3-0-β-glucopyranoside and cyanidin-3-0-β-glucopyranodise were more potent than ellagic acid, with IC_{50} of 3.0, 2.0, and 16 μg/mL, respectively. When comparing the antioxidative properties of the specific PJ anthocyanins, the anthocyanins pelargonidin-3-0-β-glucopyranoside and cyanidin-3,5-di-0-β-glucopyranoside were less potent than the other two (IC_{50} of 13 μg/mL vs. 2 to 3 μg/mL). A similar pattern was noted for the free-radical scavenging capabilities of the above PJ fractions. PJ administration to atherosclerotic patients and mice was shown to significantly increase PON1 activity.[53-55] In order to analyze which of the PJ fractions is responsible for the effect on PON1, we incubated the PJ fractions (1 μmol/L) with human serum. Cyanidin-3,5-di-0-β-glucopyranoside and pelargonidin-3-0-β-glucopyranoside significantly increased PON1 activity by 20%, whereas ellagic acid and delphinidin-3-0-β-glucopyranoside increased the activity by only 10%. Gallic acid and cyanidin-3-0-β-glucopyranodise did not affect PON1 activity.

The PJ hydrolyzable tannins are punicalin, pedunculagin, punicalagin, and gallagic and ellagic acidesters of glucose.[65] We demonstrated that punicalagin inhibited copper ion-induced LDL oxidation and this effect could be related to its capability to scavenge free radicals. Punicalagin also reduced macrophage oxidative stress by up to 90%, and the ability of the cells to oxidize LDL by up to 40%.

2.6   ANTIOXIDATIVE EFFECTS OF POMEGRANATE BYPRODUCTS

Pomegranate husk was extracted and a product of total pomegranate tannins (TPT) containing 85% punicalagin, 1.3% ellagic acid, and a small amount of ellagic acid glycosides was purified by Dr. Navindra P. Seeram from the laboratory of Dr. David Heber at UCLA, USA.[66] On a similar polyphenol weight basis (10 µg/mL), TPT was more potent than vitamin E and PJ when analyzed as a free-radical scavenger. TPT reduces the absorbance in the DPPH assay by 75% in comparison to 62% and 48% reduction obtained by a similar total polyphenol content of PJ or vitamin E, respectively. Copper ion-induced and AAPH-induced LDL oxidation were both dose-dependently inhibited by TPT with an IC_{50} of 2.1 µg/mL and 1.4 µg/mL, respectively. Macrophage oxidative status was also substantially decreased by 50% using 80 µmol/mL of TPT polyphenols. At a similar concentration, punicalagin was found to be even more potent than TPT in all the above assays.

The antioxidative properties of pomegranate polyphenol extract powder, as well as pomegranate fiber powder, were also analyzed. The pomegranate fiber powder polyphenols concentration was 8 times lower compared to the polyphenols in the pomegranate extract powder (200 ± 6 nmol/mg vs. 1580 ± 138 nmol/mg, respectively). The pomegranate extract was significantly more potent than the fiber powder in both scavenging of free radicals and in reducing macrophage oxidative stress.

Recently, we have shown[67] that pomegranate byproduct, like PJ, significantly decreased macrophage oxidative state, as well as the extent of the cellular uptake of oxidized LDL by J-774 A.1 macrophages.

2.7   CONCLUSIONS

1. All the pomegranate *tree* parts contain polyphenols and possess antioxidant activities. The tannins from bark and stem were most potent.
2. All the pomegranate *fruit* parts contain polyphenols and possess antioxidant activity. The fruit membrane anthocyanins and tannins were the most potent fruit antioxidants.
3. PJ contains the highest concentration of total polyphenols, in comparison to other fruit juices studied. The PJ polyphenols were found to be the most powerful antioxidants.
4. PJ-derived gallic acid, as well as ellagic acid and its unique tannins and anthocyanins, inhibit LDL oxidation as a result of their free-radical scavenging and metal ion chelation properties.
5. Pomegranate by-products (husk tannins) exhibit potent antioxidative effects against LDL oxidation.

Figure 2.3 summarizes the antioxidative properties of PJ.

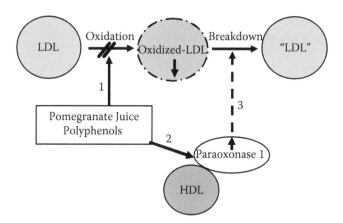

FIGURE 2.3 The antioxidative properties of pomegranate juice (PJ) polyphenols. PJ polyphenols inhibit copper ion-induced LDL oxidation, and thus reduce oxidized LDL content, 1. PJ polyphenols also increase the activity of serum HDL-associated paraoxonase 1, 2. PON1 can in turn hydrolyze lipid peroxides in oxidized LDL and convert them to a less atherogenic particle ["LDL," 3], leading to a further reduction in oxidized LDL content.

REFERENCES

1. Vaya, J. and Aviram, M., Nutritional antioxidants: mechanisms of action, analyses of activities and medical applications, *Curr. Med. Chem. Imm. Endoc. Metab. Agents,* 1, 99, 2001.
2. Kelawala, N.S. and Ananthanrayan, L., Antioxidant activity of selected foodstuffs, *Inter. J. Food and Nut.,* 55, 511, 2004.
3. Halvorsen, B.L. et al., A systematic screening of total antioxidants in dietary plants, *J. Nutr.,* 132, 461, 2002.
4. Burton, G.W., Joyce, A., and Ingold, K.U., Is vitamin E the only lipid-soluble, chain breaking antioxidant in human blood plasma and erythrocyte membranes?, *Arch. Biochem. Biophys.,* 221, 281, 1983.
5. Chu, Y.F., and Liu, R.H., Novel low-density lipoprotein (LDL) oxidation model: antioxidant capacity for the inhibition of LDL oxidation, *J. Agric. Food. Chem.,* 52, 6818, 2004.
6. Klouche, K. et al., Mechanism of *in-vitro* heme-induced LDL oxidation: effects of antioxidants, *Eur. J. Clin. Invest.,* 34, 619, 2004.
7. Maor, I. et al., Plasma LDL oxidation leads to its aggregation in atherosclerotic apolipoprotein E-deficient mice, *Arterioscler. Thromb. Vasc. Biol.,* 17, 2995, 1997.
8. Jacobsson, L.S. et al., Effects of alpha-tocopherol and astaxanthin on LDL oxidation and atherosclerosis in WHHL rabbits, *Atherosclerosis,* 173, 231, 2004.
9. Winklhofer-Roob, B.M. et al., Effects of vitamin E depletion/repletion on biomarkers of oxidative stress in healthy aging, *Ann. N.Y. Acad. Sci.,* 1031, 361, 2004.
10. Carpenter, K.L. et al., Oral alpha-tocopherol supplementation inhibits lipid peroxidation in established human atherosclerotic lesions, *Free. Radic. Res.,* 37,1235, 2003.
11. Sies, H. and Stahl, W., Vitamins E and C, β-carotene, and other carotenoids as antioxidants, *Am. J. Clin. Nutr.,* 62 (6 Suppl), 1315S, 1995.

12. Stahl, W. and Sies, H., Antioxidant defense: vitamins E and C and carotenoids, *Diabetes,* 2, S14, 1997.
13. Krinsky, N.I., Carotenoids as antioxidants, *Nutrition,* 17, 815, 2001.
14. Fuhrman, B. et al., Tomato's lycopene and β-carotene inhibit low density lipoprotein oxidation and this effect depends on the lipoprotein vitamin E content, *Nutr. Metab. Cardiovasc. Dis.,* 7, 433, 1997.
15. Fuhrman, B. et al., Lycopene synergistically inhibits LDL oxidation in combination with vitamin E, glabridin, rosmarinic acid, carnosic acid, or garlic, *Antiox. Redox. Signal.,* 2, 491, 2000.
16. Linseisen, J. et al., Effect of single oral dose of antioxidant mixture (vitamin E, carotenoids) on the formation of cholesterol oxidation products after *ex vivo* LDL oxidation in humans, *Eur. J. Med. Res.,* 3, 5, 1998.
17. Upritchard, J.E. et al., Spread supplementation with moderate doses of vitamin E and caroteniods reduces lipid peroxidation in healthy, nonsmoking adults, *Am. J. Clin. Nutr.,* 78, 985, 2003.
18. Rao, A.V., Lycopene, tomatoes, and the prevention of coronary heart disease, *Exp. Biol. Med.,* 227, 908, 2002.
19. Winkel-Shirley, B., Flavonoid biosynthesis. A colorful model for genetics, biochemistry, cell biology, and biotechnology, *Plant. Physiol.,* 126, 485, 2001.
20. Rice-Evans, C.A., Miller, N.J., and Paganga, G., Structure-antioxidant activity relationships of flavonoids and phenolic acids, *Free. Radic. Biol. Med.,* 20, 933, 1996.
21. Van Acker, SABE. et al., Structural aspects of antioxidants activity of flavonoids, *Free. Radic. Biol. Med.,* 20, 331, 1996.
22. Aviram, M. and Fuhrman, B., Effects of flavonoids on the oxidation of LDL and atherosclerosis, in *Flavonoids in health and disease,* 2nd ed. Rice-Evans, C. and Packer, L., eds., Marcel Dekker, New York, 2003, 165.
23. Vaya, J. et al., Inhibition of LDL oxidation by flavonoids in relation to their structure and calculated enthalpy, *Phytochemistry,* 62, 89, 2003.
24. Yu, J. et al., Antioxidant activity of citrus limonoids, flavonoids, and coumarins, *J. Agric. Food. Chem.,* 53, 2009, 2005.
25. Yuting, C. et al., Flavonoids as superoxide scavengers and antioxidants, *Free. Radic. Biol. Med.,* 9, 19, 1999.
26. Morel, I. et al., Antioxidant and iron-chelating activities of the flavonoids catechin, quercetin and diosmetin on iron-loaded rat hepatocyte cultures, *Biochem. Pharmacol.,* 45, 13, 1993.
27. Morel, I. et al., Role of flavonoids and iron chelation in antioxidant action, *Methods. Enzymol.,* 234, 437, 1994.
28. Fuhrman, B. and Aviram, M., Preservation of paraoxonase activity by wine flavonoids: possible role in protection of LDL from lipid peroxidation, *Ann. N.Y. Acad. Sci.,* 957, 321, 2002.
29. Aviram, M. et al., Human serum paraoxonase (PON 1) is inactivated by oxidized low density lipoprotein and preserved by antioxidants, *Free. Radic. Biol. Med.,* 26, 904, 1999.
30. Aviram, M. and Fuhrman, B., Polyphenolic flavonoids inhibit macrophage-mediated oxidation of LDL and attenuate atherogenesis, *Atherosclerosis,* 137, Suppl, S45, 1998.
31. Vaya, J., Belinky, P.A., and Aviram, M., Antioxidant constituents from licorice roots: isolation, structure elucidation and antioxidative capacity toward LDL oxidation, *Free. Radic. Biol. Med.,* 23, 302, 1997.

32. Fuhrman, B. et al., Licorice extract and its major polyphenol glabridin protect low-density lipoprotein against lipid peroxidation: *in vitro* and *ex vivo* studies in humans and in atherosclerotic apolipoprotein E-deficient mice, *Am. J. Clin. Nutr.*, 66, 267, 1997.

33. Belinky, P.A. et al., The antioxidative effects of the isoflavan glabridin on endogenous constituents of LDL during its oxidation, *Atherosclerosis*, 137, 49, 1998.

34. Fuhrman, B. et al., Antiatherosclerotic effects of licorice extract supplementation on hypercholesterolemic patients: increased resistance of LDL to atherogenic modifications, reduced plasma lipid levels, and decreased systolic blood pressure, *Nutrition*, 18, 268, 2002.

35. Rosenblat, M. et al., Macrophage enrichment with the isoflavan glabridin inhibits NADPH oxidase-induced cell-mediated oxidation of low density lipoprotein. A possible role for protein kinase C, *J. Biol. Chem.*, 14, 274, 13790, 1999.

36. Hou, L. et al., Inhibition of human low density lipoprotein oxidation by flavonols and their glycosides, *Chem. Phys. Lipids*, 129, 209, 2004.

37. Faustino, R.S. et al., Differential antioxidant properties of red wine in water soluble and lipid soluble peroxyl radical generating systems, *Mol. Cell. Biochem.*, 263, 211, 2004.

38. Filipe, P. et al., Anti- and pro-oxidant effects of quercetin in copper-induced low density lipoprotein oxidation. Quercetin as an effective antioxidant against pro-oxidant effects of urate, *Eur. J. Biochem.*, 27, 1991, 2004.

39. Fuhrman, B. et al., White wine with red wine-like properties: increased extraction of grape skin polyphenols improves the antioxidant capacity of the derived white wine. *J. Agric. Food. Chem.*, 49, 3164, 2001.

40. Fuhrman, B., Lavy, A., and Aviram, M., Consumption of red wine with meals reduces the susceptibility of human plasma and low-density lipoprotein to lipid peroxidation, *Am. J. Clin. Nutr.*, 61, 549, 1995.

41. Aviram, M., Hayek, T., and Fuhrman, B., Red wine consumption inhibits LDL oxidation and aggregation in humans and in atherosclerotic mice, *Biofactors*, 6, 415, 1997.

42. Howard, A. et al., Red wine consumption and inhibition of LDL oxidation: what are the important components? *Med. Hypotheses*, 59, 101, 2002.

43. Vigna, G.B. et al., Effect of a standardized grape seed extract on low-density lipoprotein susceptibility to oxidation in heavy smokers, *Metabolism*, 52, 1250, 2003.

44. Fuhrman, B. et al., Grape powder polyphenols attenuate atherosclerosis development in apolipoprotein E deficient (E^0) mice and reduce macrophage atherogenicity, *J. Nutr.*, 135, 722, 2005.

45. Hayek, T. et al., Reduced progression of atherosclerosis in apolipoprotein E-deficient mice following consumption of red wine, or its polyphenols quercetin or catechin, is associated with reduced susceptibility of LDL to oxidation and aggregation, *Arterioscler. Thromb. Vasc. Biol.*, 17, 2744, 1997.

46. Aviram, M. and Fuhrman, B., Wine flavonoids protect against LDL oxidation and atherosclerosis, *Amm. N.Y. Acad. Sci.*, 957, 146, 2002.

47. Stocker, R. and O'Halloran, R.A., Dealchoholized red wine decreases atherosclerosis in apolipoprotein E gene-deficient mice independently of inhibition of lipid peroxidation in the artery wall, *Am. J. Clin. Nutr.*, 79, 123, 2004.

48. Cemeroglu, B., Artik, N., and Erbas, S., Extraction and composition of pomegranate juice, *Fluessiges. Obst.*, 59, 335, 1992.

49. El-Nemr, S.E., Ismail, I.A., and Ragab, M., Chemical composition of juice and seeds of pomegranate fruit, *Nahrung*, 34, 601, 1991.

50. Melgarejo, P., Salazar, D.M., and Artes, F., Organic acids and sugars composition of harvested pomegranate fruits, *Eur. Food. Res. Technol.,* 211, 185, 2000.
51. NarrBen, C., Ayed, N., and Metche, M., Quantitative determination of the polyphenolic content of pomegranate peel, *Z lebensm. Unters. Forsch.,* 203, 374, 1998.
52. El-Toumy, S.A.A. and Marzouk, M.S.A., A new ellagic acid glycoside from *Punica granatum* L, *Polyphenols. Commun.,* Freising-Weihenstephan (Germany), September 11–15, 127, 2000.
53. Aviram, M. et al., Pomegranate juice consumption reduces oxidative stress, atherogenic modifications to LDL, and platelet aggregation: studies in humans and in atherosclerotic apolipoprotein E-deficient mice, *Am. J. Clin. Nutr.,* 71, 1062, 2000.
54. Aviram, M. et al., Pomegranate juice consumption for 3 years by patients with carotid artery stenosis reduces common carotid intima-media thickness, blood pressure and LDL oxidation, *Clin. Nutr.,* 23, 423, 2004.
55. Kaplan, M. et al., Pomegranate juice supplementation to atherosclerotic mice reduces macrophage lipid peroxidation, cellular cholesterol accumulation and development of atherosclerosis, *J. Nutr.,* 131, 2082, 2001.
56. De Nigris, F. et al., Beneficial effects of pomegranate juice on oxidation-sensitive genes and endothelial nitric oxide synthase activity at sites of perturbed shear stress, *Proc. Natl. Acad. Sci.,* 102, 4896, 2005.
57. El-Toumy, S.A. and Rauwald, H.W. Two ellagitannins from *Punica granatum* heartwood, *Phytochemistry,* 61, 971, 2002.
58. Noda, Y. et al., Antioxidant activities of pomegranate fruit extract and its anthocyanidins: delphinidin, cyanidin, and pelargonidin, *J. Agric. Food. Chem.,* 50, 166, 2002.
59. Singh, R.P., Chidambara Murthy, K.N., and Jayaprakasha, G.K., Studies on the antioxidant activity of pomegranate *(Punica granatum)* peel and seed extracts using *in vitro* models, *J. Agric. Food. Chem.,* 50, 81, 2002.
60. Schubert, S.Y., Lansky, E.P., and Neeman, I., Antioxidant and eicosanoids enzyme inhibition properties of pomegranate seed oil and fermented juice flavonoids, *J. Ethnopharmacol.,* 66, 11, 1999.
61. Wang, R.F. et al., Bioactive compounds from the seeds of *Punica Granatum* (pomegranate), *J. Nat. Prod.,* 67, 2096, 2004.
62. Plumb, G.W. et al., Antioxidant properties of gallocatechin and prodelphinidins from pomegranate peel, *Redox. Rep.,* 7, 41, 2002.
63. Gil, M.I. et al., Antioxidant activity of pomegranate juice and its relationship with phenolic composition and processing, *J. Agric. Food. Chem.,* 48, 4581, 2000.
64. Schubert, S.Y., Neeman, I., and Resnick, N., A novel mechanism for the inhibition of NF-kB activation in vascular endothelial cells by natural antioxidants, *FASEB. J.,* 16, 1931, 2002.
65. Afaq, F. et al., Pomegranate fruit extract modulates UV-B-mediated phosphorylation of mitogen-activated protein kinases and activation of nuclear factor kappa B in normal human epidermal keratinocytes paragraph sign, *Photochem. Photobiol.,* 81, 38, 2005.
66. Seeram, N. et al., Rapid large scale purification of ellagitannins from pomegranate husk, a by-product of the commercial juice industry, *Separation and Purification Technology,* 41, 49, 2005.
67. Rosenblat, M., Volkova, N., and Aviram, M., Pomegranate byproduct administration to apolipoprotein E-deficient mice attenuates atherosclerosis development as a result of decreased macrophage oxidative stress and reduced cellular uptake of oxidized low density lipoprotein. *J. Agric. Food Chem.* 54, 1928, 2006.

3 Bioavailability of Pomegranate Polyphenols

Francisco A. Tomás-Barberán, Navindra P. Seeram, and Juan Carlos Espín

CONTENTS

3.1 INTRODUCTION

Polyphenols are relevant constituents regarding the organoleptic properties of pomegranate arils and juice as they are responsible for the distinctive red pigmentation and provide a mild astringency that is characteristic of pomegranate flavor. The juice obtained from these arils contains anthocyanins (delphinidin, cyanidin, and pelargonidin 3-glucosides and 3,5-diglucosides), ellagic acid glycosides (ellagic acid glucoside, arabinoside, and rhamnoside), free ellagic acid (Figure 3.1), ellagitannins (several punicalagin isomers, punicalin, and some punicalagin polymeric forms) (Figure 3.2), and gallotannins. The husk and fruit membranes contain mainly ellagitannins that are water soluble (punicalagins),[1] and small amounts of procyanidins (prodelphinidins and gallocatechin).[2] Anthocyanins are also present in the skin, although the delphinidin derivatives are not generally observed and the cyanidin and pelargonidin derivatives coincide with those found in the juice.[3] During industrial

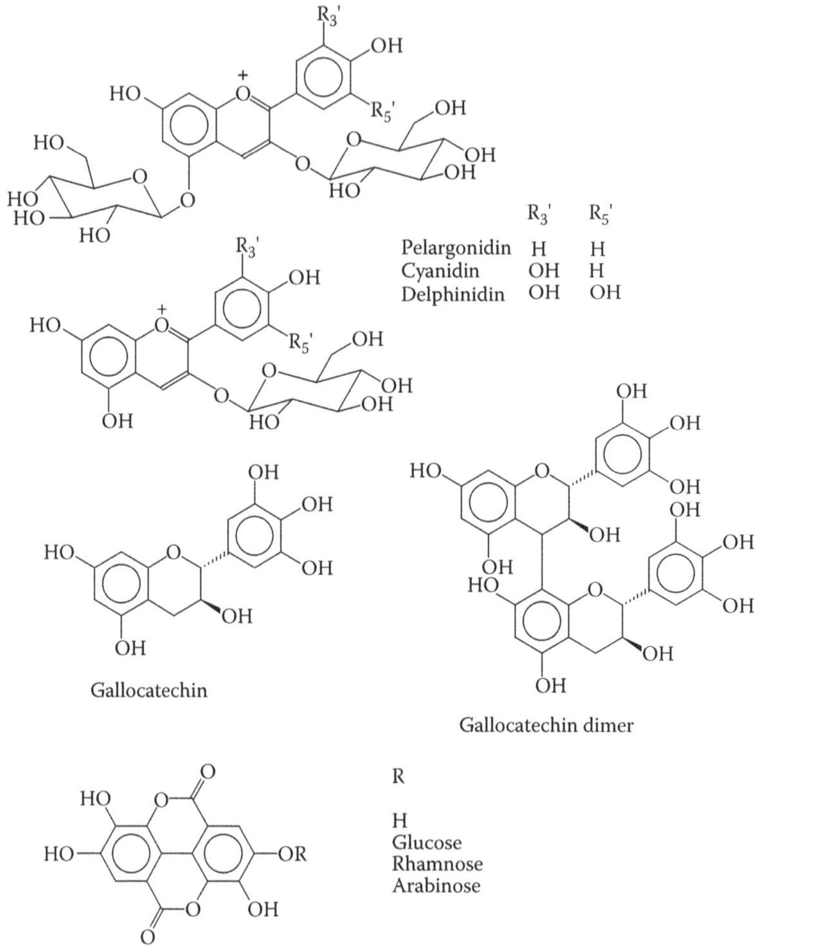

FIGURE 3.1 Anthocyanins, proanthocyanidins, and ellagic acid glycosides from pomegranate.

processing, the technological treatment allows the release of water-soluble husk punicalagins into the juice, which has been related to the outstanding antioxidant activity observed in commercial pomegranate juice.[1]

The biological activity attributed to pomegranates and pomegranate juice has been related to the content in antioxidant phenolic compounds.[4] It is therefore essential to investigate the bioavailability and metabolism of the pomegranate polyphenols *in vivo,* as well as the tissue distribution and relevant concentrations of the corresponding metabolites, in order to understand their role in health promotion. In addition, it is important to evaluate the potential toxicity of some of these phenolics, as punicalagin and related compounds have been associated to kidney toxicity in cattle after ingestion of large amounts of grass containing these kinds of compounds.[5]

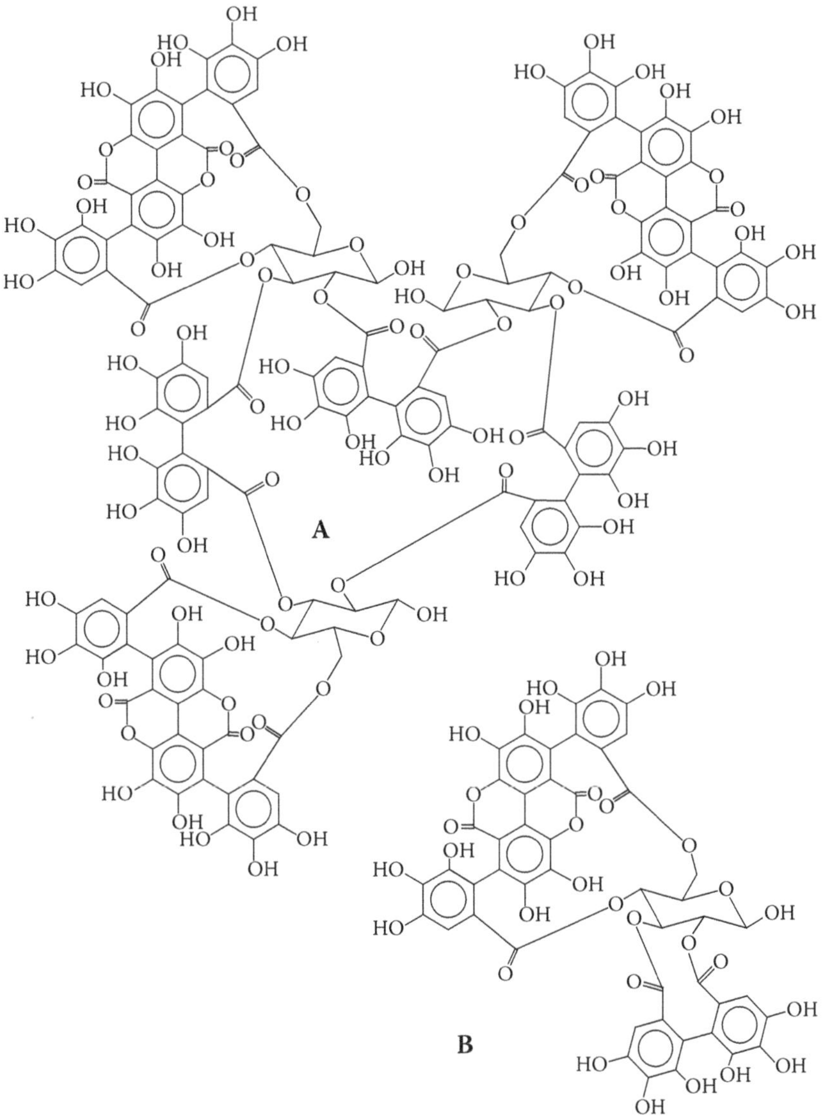

FIGURE 3.2 Ellagitannins from pomegranate. (A) Punicalagin trimer; (B) Punicalagin.

3.2 ANIMAL AND CELL BIOAVAILABILITY STUDIES

As a general rule, before studying the bioavailability of polyphenols in humans, animal studies are recommended as larger intakes can be attained and toxicity effects can be ruled out. The bioavailability of pomegranate phenolics has been directly or indirectly evaluated in animals and results on ellagic acid and ellagitannins, anthocyanins, and proanthocyanidins are available.

3.2.1 Ellagic Acid and Ellagitannins

In an early study, the bioavailability of ellagic acid in the mouse following oral administration by gavage was evaluated.[6] This study provided 0.3 μg/g body weight (the equivalent for a human of 70 kg of 21-mg intake) and free ellagic acid and conjugates were observed in urine, bile, and blood. Sulphate ester, glucuronide, and glutathione conjugates of ellagic acid were present in urine, bile, and blood. In this case it was observed that absorption of 3H-ellagic acid occurred mostly within two hours of oral administration. Levels in blood, bile, and tissues were low, and almost all of the observed dose was excreted in urine. More than half of the administered 3H-ellagic acid remained in the gastrointestinal tract after 24 hours. Approximately 19% was excreted with faeces and 22% in urine at 24 hours.

A rapid absorption and metabolism of ellagic acid is reported by Doyle and Griffiths in rats,[7] and this does not agree with the observations of Smart et al. for CD-1 mice.[8] Doyle and Griffiths found two metabolites in faeces and urine of the rat. One was 3,8-dihydroxy-6H-dibenzo-[b,d]pyran-6-one (urolithin-A) and the other was not identified.[7] Both were reported to be of microfloral origin since neither was found in germ-free animals. Unchanged ellagic acid was not detected in urine or faeces of normal rats. In the work by Smart et al.[8] they found low to nondetectable levels of ellagic acid in blood, lungs, and liver of CD-1 mice after oral administration, which was interpreted to indicate poor absorption and rapid elimination of the compound in these animals.

The disposition described by Teel and Martin[6] agrees with the report of Smart et al.,[8] both in mice, suggesting that the poor absorption of ellagic acid may prevent tissues from attaining sufficiently high concentrations for it to be efficacious as an *in vivo* anticarcinogenic agent. The poor absorption of ellagic acid is supported by a report of the presence of ellagic acid calculi in the gastrointestinal tract of monkeys and goats whose diet naturally contains ellagic acid.[9] The poor absorption may be caused by several factors including the ionization of ellagic acid at physiological pH and the formation of poorly soluble complexes with magnesium and calcium ions. In addition, the extensive binding of ellagic acid to intestinal epithelium could be also involved in its poor absorption.[10]

Concerning pomegranate ellagitannins, a recent work studied the bioavailability of pomegranate husk ellagitannins in the rat.[11] These ellagitannins are essentially the same as those found in the commercial juice.[1] The rats were given 6% of their diet as pomegranate ellagitannins from the fruit husk and the experiment was also used to evaluate tissue distribution and toxicity. Values around 3 to 6% of the ingested punicalagin were excreted as metabolites in faeces and urine. In faeces, punicalagin is transformed to hydrolysis products and to hydroxy-6H-dibenzo-[b,d]-pyran-6-one derivatives (Figure 3.3) by the rat colonic microflora. In plasma, punicalagin was detected at concentrations around 30 μg/mL (0.028 μM). The absorption of intact punicalagin in rats and its detection in plasma is especially relevant as this is one of the largest polyphenols (1084 a.m.u.) that has been reported to be absorbed.[11,12] Glucuronides and methyl ether conjugates of ellagic acid were also detected in this study. In urine the main metabolites observed were the monohydroxy, dihydroxy, and trihydroxy-6H-dibenzo-[b,d]-pyran-6-one derivatives, as aglycones or glucu-ronide conjugates.[11] It was concluded that as only 3 to 6% of the ingested punicalagin was detected as such or as metabolites in urine and faeces, the majority of this ellagitannin has to be converted to undetectable metabolites (i.e., CO_2) or accumu-lated in nonanalysed tissues. However, only traces of punicalagin metabolites were detected in liver or kidney and no intact punicalagin, ellagic acid, or other derived metabolites were detected in lung, brain, or heart.[11,13]

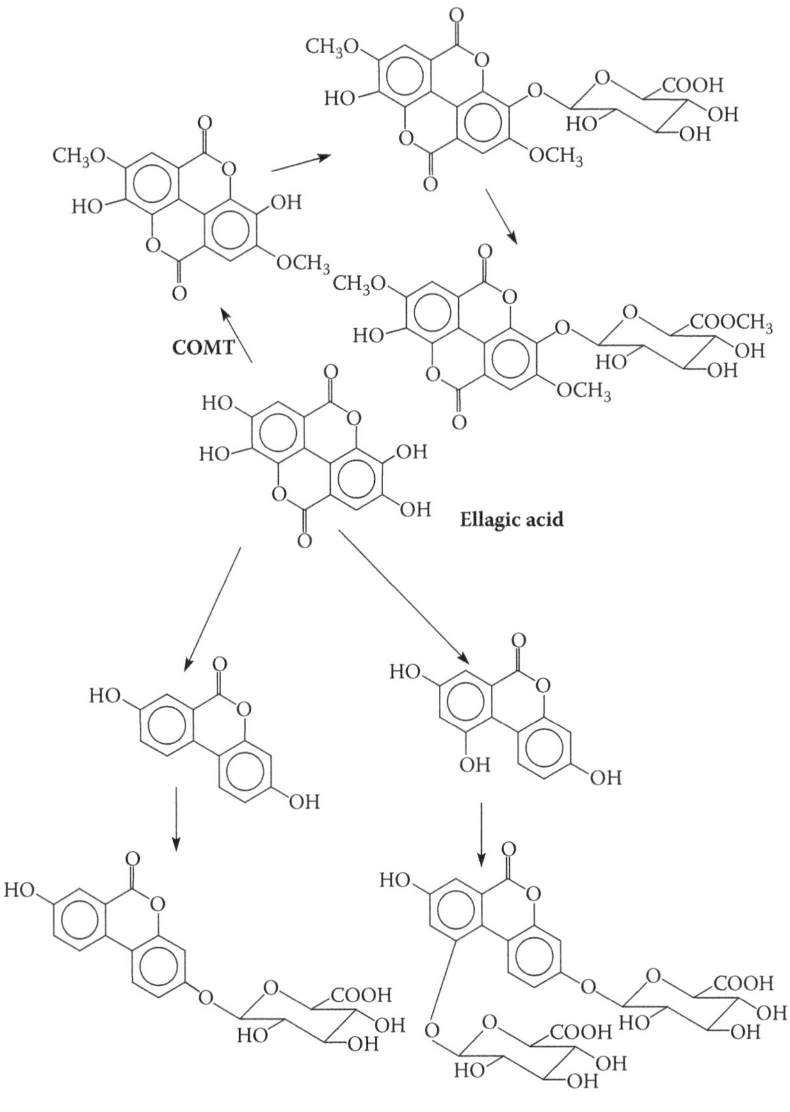

FIGURE 3.3 Punicalagin metabolites in rat.

In this experiment, in addition to the very small amounts of punicalagin, the pomegranate ellagitannin metabolites detected in plasma included trihydroxy-6H-dibenzopyran 6-one diglucuronide and dihydroxy-6H-dibenzopyran 6-one glucuronide, gallagic acid, dimethyl-ellagic acid glucuronide, dimethyl ellagic acid glucuronide methyl ester, and very small amounts of ellagic acid (Figure 3.3). As ellagic acid has two orthodihydroxy groups it can be expected that, in the liver, via the activity of COMT (catechol o-methyl transferase), one methyl ether can be introduced per dihydroxyl grouping. These metabolites show UV spectra nearly identical to that of free ellagic acid, and were further conjugated with glucuronic acid to increase water solubility and facilitate the excretion.[11]

A relevant factor that could explain low recovery of these phenolics or their metabolites is that these types of compounds bind to proteins in the intestinal cells, in plasma, or in the faecal material, and their extraction for analysis from these tissues and biological fluids can underestimate the actual metabolites, as the recovery can be difficult with the usual extraction procedures.[10]

There are only a few reports regarding cell models to explore ellagic acid uptake and metabolism. Whitley et al. reported a high accumulation of ellagic acid in Caco-2 cells (uptake through Caco-2 monolayer epithelium), indicating a facile absorptive transport across the apical membrane.[10] As much as 93% of the cellular ellagic acid was irreversibly bound to macromolecules (proteins and DNA). Thus ellagic acid appears to accumulate in the epithelial cells of the aerodigestive tract.[10]

In a recent study, the metabolism of punicalagin and ellagic acid by Caco-2 cells was reported.[14] This study showed that punicalagin is hydrolyzed in the cell medium to yield ellagic acid, which enters into Caco-2 cells. The first ellagic-acid-derived metabolite produced by these cells was dimethyl-ellagic, which involved the active participation of COMT in this metabolism. Afterwards, Caco-2 cells conjugated glucuronic acid to produce the corresponding dimethyl-ellagic acid glucuronide, which was the most abundant metabolite produced (Figure 3.3). Other metabolites detected in smaller amounts included two isomers of dimethyl-ellagic acid sulphates. All these metabolites were detected in both medium and inside the cells.[14]

Other studies to evaluate the bioavailability and metabolism of ellagitannins in model animals are currently underway using the pig as a model and acorn ellagitannins as the phenolic compounds to be evaluated. Although this study is not yet concluded, we can anticipate that the main metabolites observed in urine and plasma are those related to the transformation of ellagic acid into the hydroxyl-dibenzo-[b,d]pyran-6-one derivatives (urolithin A and B as glucuronides and in small amounts as aglycones) (Figure 3. 4).

The biodistribution of ellagic acid has been previously studied in mice as a function of dose and time after gavage of ellagic acid.[15] The levels of ellagic acid in the lung were directly proportional to the dose of ellagic acid between 0.2 and 2.0 mmol. The maximum level of ellagic acid, corresponding to 21.3 nmol/g, was observed 30 minutes after gavage with 2.0 mmol of ellagic acid/kg body weight, which corresponds to only 70 ppm of the administered dose. The levels in liver tissues were 10-fold lower and reached a maximum 30 minutes after gavage. At this interval, the blood level of ellagic acid was 1 nmol/ml. Interestingly, the inclusion of ellagic acid in cyclodextrin doubles the level of ellagic acid in lung tissues.[15]

Quantification of ellagic acid, the principal bioactive component of a pomegranate leaf extract, in rat plasma following oral administration of pomegranate leaf extract, has also been determined.[16] The concentration–time profile was fitted with an open two-compartment system with lag time, and its maximum concentration of ellagic acid in plasma was 213 ng/ml only 0.55 h after oral administration of 0.8 g extract per kg body weight. The pharmacokinetic profile indicates that ellagic acid has poor absorption (in accordance with the above study carried out in mice) and rapid elimination after the oral administration of the pomegranate leaf extract. This also indicated that part of ellagic acid was absorbed in the stomach.[16] The absorption of ellagic acid from this pomegranate leaf extract was reported to be higher than those reported with the same animal model when pure ellagic acid was ingested, and the authors suggested that other compounds present in the extract could facilitate the absorption, probably by increasing the ellagic acid solubility in the gastrointestinal fluids or by inhibiting the intestinal microorganisms that are responsible for ellagic acid metabolism. This latter suggestion anticipated the critical role of the intestinal flora in the metabolism of ellagic acid to yield urolithin metabolites.

Free gallic acid, a compound that can be present in pomegranate extracts in a free state that can be obtained after the release from gallotannins, has one of the largest bioavailabilities of the food phenolic compounds. Values of urinary excretion close to 40% of the ingested amounts have been reported.[12]

3.2.2 ANTHOCYANINS

Although there is no direct evidence of the bioavailability of pomegranate anthocyanins, there are some studies regarding the bioavailability of anthocyanins that should be mentioned.

McGhie et al.[17] investigated the absorption of 15 anthocyanins with different aglycone structures and conjugated sugars extracted from blueberry, boysenberry, black raspberry, and black currant in both humans and rats. Intact and unmetabolized anthocyanins were detected in urine of rats and humans following dosing for all molecular structures investigated, thus demonstrating that anthocyanins with diverse molecular structure and from different dietary sources are bioavailable at diet-relevant dosage rates.[17] In addition, the relative concentrations of anthocyanins detected in urine following dosing varied, indicating that differences in bioavailability are due to variations in chemical structure. These results suggested that the nature of the sugar conjugate and the phenolic aglycone are both important determinants of anthocyanin absorption and excretion in rats and humans. In fact the bioavailability of pelargonidin glycosides is reported to be higher than those of other related anthocyanins of the cyanidin and delphinidin type.[12,17]

Other recent studies have reported the absorption and bioavailability of anthocyanins in different animal models. Wu et al.[18] showed the different absorption and metabolism of pelargonidin and cyanidin after marionberry consumption in pigs. The authors fed weaning pigs (7.9 ± 1.7 kg) with a freeze-dried powder of marionberry by stomach tube. Marionberry contains two anthocyanins that are also present in pomegranate juice, namely cyanidin 3-glucoside and pelargonidin 3-glucoside (in addition to cyanidin rutinoside). In the urine, the original anthocyanins and different

metabolites were identified and quantified. The main metabolites were glucu-
ronidated or methylated forms of the original anthocyanins. Total recovery of the
original anthocyanins plus their related metabolites was $0.087 \pm 0.034\%$ for cyanidin
3-glucoside, and $0.583 \pm 0.229\%$ for pelargonidin 3-glucoside. In pig plasma, the
original cyanidin 3-glucoside, and trace amounts of one metabolite, cyanidin monoglucuronide, were detected. The plasma concentration:dose ratio of cyanidin
3-rutinoside was higher than that of its 3-glucoside. In accordance with McGhie
et al.,[17] the authors suggested that the nature of the sugar conjugate and the phenolic
aglycone are both important determinants of anthocyanin absorption and excretion.
Cyanidin 3-glucoside and the 3-rutinoside had similar apparent excretion rates relative to dose, whereas pelargonidin 3-glucoside had a much higher total urinary
excretion than cyanidin-based anthocyanins. Most of cyanidin-3-glucoside and
pelargonidin-3-glucoside were excreted in the form of metabolites.

Recently, Matuschek et al.[19] have reported that the jejunum is the main site of
absorption for anthocyanins in mice using mouse duodenum, jejunum, ileum, and
colon. Minor absorption occurred with duodenal tissue with no absorption recorded
when tissues from ileum or colon were used.

In another study, Talavera et al.[20] fed rats with a blackberry (anthocyanin-
enriched) diet for 15 days. Different anthocyanin metabolites were detected in the
stomach (native blackberry anthocyanins, or cyanidin 3-O-glucoside and cyanidin
3-O-pentoside), while in other organs (jejunum, liver, and kidney) native and methylated anthocyanins as well as conjugated anthocyanidins (cyanidin and peonidin
monoglucuronides) were identified. Proportions of anthocyanin derivatives differed
according to the organ considered, with the liver presenting the highest proportion
of methylated forms. Jejunum and plasma also contained aglycone forms. In the
brain, the total anthocyanin content (blackberry anthocyanins and peonidin 3-O-
glucoside) reached 0.25 ± 0.05 nmol/g of tissue (n = 6). The urinary excretion of
total anthocyanins was low ($0.19 \pm 0.02\%$ of the ingested amount). Thus, organs of
the digestive area indicated a metabolic pathway of anthocyanins with enzymatic
conversions (methylation or glucurono-conjugation).[20] Moreover, these authors reported
that following consumption of an anthocyanin-rich diet, anthocyanins enter the brain.
This finding was also described by Andres-Lacueva et al.[21] who fed rats with blueberry
for 8 to 10 weeks to analyze anthocyanins and derived metabolites in different brain
regions. These authors found several anthocyanins (cyanidin-3-O-galactoside, cyanidin-
3-O-glucoside, cyanidin-3-O-arabinoside, malvidin-3-O-galactoside, malvidin-3-O-
glucoside, malvidin-3-O-arabinoside, peonidin-3-O-arabinoside, and delphinidin-
3-O-galactoside) in the cerebellum, cortex, hippocampus, or striatum of the rats. In
fact, these findings were the first to suggest that polyphenolic compounds were able
to cross the blood brain barrier and localize in various brain regions important for
learning and memory.[21]

In summary, these studies indicate that the anthocyanin 3-glucosides present in
pomegranate juice can be absorbed intact and excreted as such in urine, but at very
low rates (from 0.1 to 0.6% of the intake depending on the structure), and that the
original anthocyanins are partly metabolized to methyl ether derivatives or glucuronides before excretion. These pigments, or their metabolites, can even reach tissues
such as the brain, although at very low levels. Nothing, however, is known about

the anthocyanin 3,5-diglucosides present in pomegranate juice. Anthocyanins are largely metabolized in the large intestine by the colon microflora to give simpler phenolics (phenyl acetic and phenyl propionic derivatives) that can then be absorbed and exert biological actions.[12]

3.2.3 PROCYANIDINS

Currently there is no direct evidence of the bioavailability of pomegranate procyanidins and flavan-3-ols. The bioavailability of procyanidins and flavan-3-ols is quite variable depending on the structure of the flavanol hydroxylation pattern and on the degree of polymerization.[12] Catechin and epicatechin belong to the most bioavailable polyphenols described so far, with values reported from 1% up to 30% urine excretion of the ingested amounts. In the case of epigallocatechin, a flavan-3-ol reported to be present in pomegranate husk,[2] a urinary excretion of 11.1% of the ingested value has been reported.[12] The bioavailability of the galloylated derivatives such as epigallocatechin gallate is much lower, with values of excretion generally below 0.1% of the ingested amount.[12] In the case of oligomeric and polymeric procyanidins, current studies available are inconclusive. Recently, Tsang et al.[22] fed rats with grape-seed extract (GSE) containing (+)-catechin, (−)-epicatechin, and dimers, trimers, tetramers, and polymeric procyanidins. Small amounts of the GSE flavan-3-ols moved out of the stomach and into the duodenum/jejunum, and to a greater extent the ileum, 1 h after ingestion, and into the caecum after 2 h, with relatively small amounts being detected in the colon after 3 h. The GI tract contained the parent GSE flavan-3-ols and procyanidins with only trace amounts of metabolites and there were no indications that proanthocyanidins were depolymerised in the GI tract releasing monomeric flavan-3-ols. Plasma contained exclusively catechin glucuronides and methylated glucuronide metabolites that were also detected in the liver and kidneys. These metabolites were also present in urine together with sulphated metabolites and low amounts of the procyanidin dimers B1, B2, B3, and B4 as well as the trimer C2 and an unknown GSE trimer. This latter is one of the very scarce studies that report the urine excretion of procyanidin dimers and trimers. The amounts of (+)-catechin and (−)-epicatechin metabolites excreted in urine relative to the quantity of the monomers ingested were 27 and 36%, respectively, after 24 h. This is similar to the levels of urinary excretion reported to occur by other investigators after feeding (−)-epicatechin to rats and provides further, albeit indirect, evidence that the procyanidin oligomers in the GSE were not depolymerised to monomers to any extent after ingestion. Unlike the distribution of anthocyanins to the brain, no convincing proof has been reported for the presence of flavan-3-ol metabolites in the brain.

3.3 TOXICITY STUDIES

High oral doses of pomegranate punicalagin (equivalent to the consumption of approximately 194 L/day of pomegranate juice by a 70-kg person, which is by far a large safety margin) were supplied to rats for 37 days in order to evaluate possible toxic or accumulative effects.[13] The effects on liver and kidney toxicity and accumulation of metabolites on these tissues were also evaluated. Five punicalagin-related metabolites were identified in liver and kidney. Two ellagic acid derivatives, gallagic acid, 3,8-dihydroxy-6H-dibenzo[b,d]pyran-6-one glucuronide, and 3,8,10-trihydroxy-6H-dibenzo[b,d]pyran-6-one were identified by HPLC-MS-MS in liver and kidney tissues of the rats.

No toxic effect was observed in eight haematological parameters, and sixteen serobiochemical parameters, including the AST, ALT, and ALP enzymes and the antioxidant enzymes glutathion peroxidase and superoxide dismutase.[13] Histopathological analyses of liver and kidney corroborated the absence of toxicity as no differences were observed between the control-rat organs and those of rats under the high pomegranate punicalagin subchronic intake.

3.4 HUMAN BIOAVAILABILITY AND PHARMACOKINETIC STUDIES

3.4.1 ELLAGIC ACID AND ELLAGITANNINS

In humans, two types of experiments have been carried out: pharmacokinetic studies, in which the absorption of pomegranate polyphenols has been evaluated during the hours following intake, and bioavailability and metabolism studies, in which the metabolites present in plasma and urine are evaluated for longer periods of time.

In one pharmacokinetic study carried out at the University of California, Los Angeles (juice obtained from "Wonderful" cultivar), after the intake of 180 mL of pomegranate juice concentrate (containing 25 mg free ellagic acid and 318 mg ellagitannins) by one volunteer, ellagic acid was detected in human plasma at a maximum concentration after 1 h postingestion (31.9 ng/mL; ca 0.1 μM) but was rapidly eliminated by 4 h.[23] In another study carried out in Murcia (Spain) (juice produced from "Mollar" cultivar), after the intake of 1 L juice by one healthy volunteer, no ellagic acid was detected in plasma during the 4 h following juice intake.[24] This could be mainly explained by the interindividual variability. Another pharmacokinetic study on the absorption of ellagic acid from black raspberry has also shown a poor (less than 1% of the ingested ellagic acid) but significant absorption of free ellagic acid during the first two hours after the intake,[25] supporting the findings of the University of California.

Larger pharmacokinetic studies with a higher volunteer number are necessary to evaluate the effect of interindividual variability and that of the amount of free ellagic acid in the juice. A recently completed pharmacokinetic study at the University of California, Los Angeles[44] using a larger number of subjects (n = 19), resulted in similar pharmacokinetic parameters for ellagic acid as observed for their previous study with one human subject.[23]

Regarding the bioavailability and metabolism studies in which the metabolites present in plasma and urine are evaluated for longer periods of time, there are three main studies to be emphasized.

One of these studies consisted of the daily intake of 1 L of pomegranate juice (containing 4.37 g/L punicalagin isomers) for 5 days by healthy volunteers.[24] Neither punicalagin nor ellagic acid present in the juice were detected in both plasma and

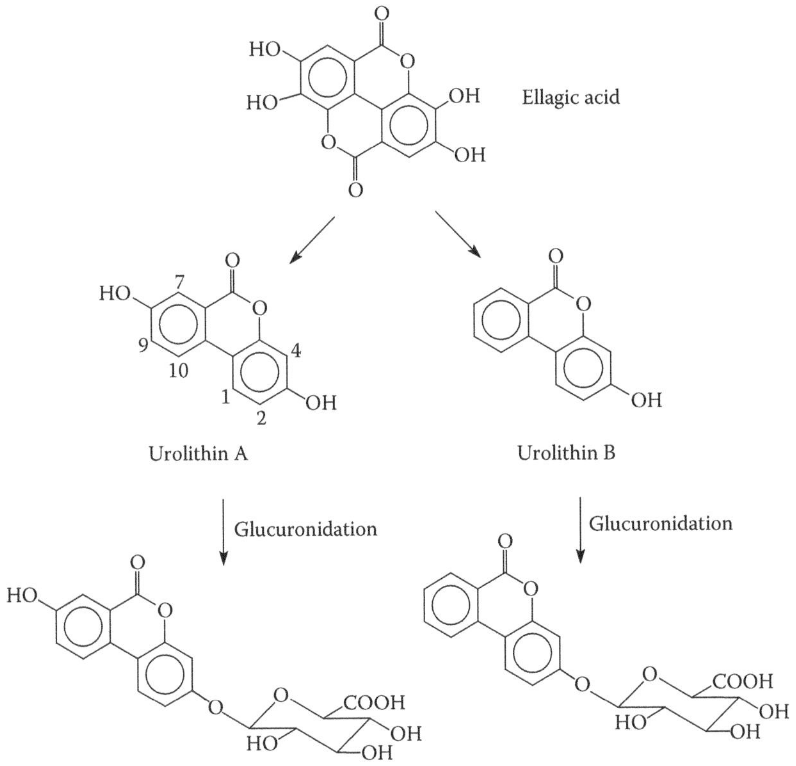

FIGURE 3.4 Punicalagin, ellagic acid, and other ellagitannin metabolites in humans.

urine. Three microbial ellagitannin-derived metabolites (urolithin derivatives) were detected: 3,8-dihydroxy-6H-dibenzo[*b,d*] pyran-6-one glucuronide, trihydroxy-6H-dibenzo[*b,d*] pyran-6-one, and 3-hydroxy-6H-dibenzo[*b,d*] pyran-6-one glucuronide (Figure 3.4). The concentration of metabolites found in plasma was quite variable, with values between 0.5 and 18.6 µM in those volunteers in which the metabolites were detected, involving a large interindividual variability. In urine, the same metabolites and their corresponding aglycones became evident after 1 day of juice consumption. Total urine excretion of metabolites ranged from 0.7 to 52.7% regarding the ingested punicalagin. In general, the metabolites associated with pomegranate juice intake coincided with those previously reported to be produced in rats after pomegranate husk intake.[11] As mentioned above, these metabolites (urolithins) were also reported to be produced by intestinal bacteria in rats after consumption of ellagic acid.[7]

Another study regarding ellagitannins bioavailability and metabolism was not carried out with pomegranate but with other ellagitannin-rich foodstuffs such as strawberry, raspberry, walnut, and oak-aged red wine.[26] These foodstuffs differ in the amount and type of ellagitannins; however, it is notable that the same metabolite (urolithin A) was detected in all the volunteers (n = 40). This led to the proposal that the microbial metabolite urolithin A[26,27] is a biomarker of human exposure to dietary ellagitannins and may be useful in intervention studies with ellagitannin-containing food products, including pomegranate. However, as mentioned above, a large interindividual variability was observed regarding the excretion of this metabolite.

The third reported study on ellagitannin bioavailability was carried out with pomegranate juice supplementation for 5 weeks on patients (n = 15) with stable chronic obstructive pulmonary disease (COPD) in a randomized, double-blind, placebo-controlled trial.[28] Again, urolithins A and B (3,8-dihydroxy-dibenzo-pyran-6-one and 3-hydroxy-dibenzo-pyran-6-one, respectively) were detected in both plasma and urine, again with a large interindividual variability.

3.4.2 ANTHOCYANINS

No bioavailability studies have been carried out on pomegranate anthocyanins in humans. However, as in the case of animal studies, there are human studies carried out with those anthocyanins present in pomegranate but obtained from other food sources.

Most of the studies on anthocyanin bioavailability in humans have been undertaken by giving the volunteers berries (whole or freeze-dried), berry extracts, or berry juices. Anthocyanin concentrations measured in plasma were usually very low (about 10 to 50 nmol/L). The mean time to reach Cmax for plasma was 1.5 h (range: 0.75 to 4 h) and 2.5 h for urine. Most studies reported low relative urinary excretions, ranging from 0.004% to 0.1% of the intake, although Lapidot et al.[29] and Felgines et al.[30] measured higher levels of anthocyanin excretion (up to 5%) after red wine or strawberry consumption. The time course of absorption was consistent with absorption in the stomach, as described for animals.[31,32] Therefore, it is noticeable that anthocyanins are usually very rapidly absorbed and eliminated. However, they are absorbed with rather poor efficiency.

There are a number of detailed examples regarding absorption and metabolism of anthocyanins in humans. For instance, in elderly women, after anthocyanin (elderberry and lowbush blueberry) intakes of 690 to 720 mg, the two major anthocyanins in elderberry, cyanidin-3-glucoside (also present in pomegranate juice) and cyanidin-3-sambubioside, as well as four metabolites (peonidin 3-glucoside, peonidin 3-sambubioside, peonidin monoglucuronide, and cyanidin 3-glucoside monoglucuronide), were identified in urine within 4 h of consumption.[33] Total elderberry anthocyanin excretion was 0.077% of intake. In women fed blueberries, urine samples contained the original anthocyanins. Reasonable correlations between blueberry and urine proportions of the different anthocyanins were obtained except for anthocyanin arabinosides. Total urinary excretion during the first 6 h was 0.004% of intake. Plasma anthocyanin levels were below detection limits using 2 mL plasma in women who consumed blueberry. This study demonstrated for the first time that *in vivo* methylation of cyanidin to peonidin and glucuronide conjugate formation occurs after people consume anthocyanins and demonstrates the low absorption and excretion of anthocyanins compared with other flavonoids.[33]

In a study to evaluate the bioavailability of strawberry anthocyanins, six healthy volunteers consumed a meal containing 200 g strawberries (providing 179 μmol pelargonidin-3-glucoside, an anthocyanin also present in pomegranate). In addition to pelargonidin-3-glucoside, five anthocyanin metabolites were identified in urine: three monoglucuronides of pelargonidin, one sulfoconjugate of pelargonidin, and pelargonidin itself.[30] Total urinary excretion of strawberry anthocyanin metabolites corresponded to 1.80 to 0.29% of pelargonidin-3-glucoside ingested. More than 80% of this excretion was related to a monoglucuronide. Four hours after the meal, more than two thirds of anthocyanin metabolites had been excreted, although urinary excretion of the metabolites continued until the end of the 24-h experiment. This study demonstrated that anthocyanins were glucuro- and sulfo-conjugated in humans and that the main metabolite of strawberry anthocyanins in human urine was a monoglucuronide of pelargonidin.[30]

It has been demonstrated that anthocyanins were detected as the original glycosides both in human plasma and urine samples.[34] The elimination of plasma anthocyanins appeared to follow first-order kinetics and most anthocyanin compounds were excreted in urine within 4 h after feeding. The current findings appear to refute assumptions that anthocyanins are not absorbed in their unchanged glycosylated forms in humans.[34]

3.4.3 PROCYANIDINS

The metabolic rate of procyanidins after consumption in humans is not yet fully understood.[35] After consumption of 2 g of high-procyanidin grape seed extract by volunteers, the plasma concentrations of procyanidin B1 reached only 10 nmol/L;[36] after consumption of 0.375 g cocoa/kg body weight, the plasma concentrations of procyanidin B2 reached only 41 nmol/L.[37] When administered in a purified form to rats, procyanidin dimer B3 was not found in the plasma.[38] However, there is much evidence regarding the relatively fast absorption and excretion of flavanols.[35] Bioavailability differs markedly among catechins. Flavanol accumulation in plasma is detected after 2 h of the ingestion with concentrations ranging from 0.1 μM to 6 μM after cocoa beverage or tea intake.[35] For these reasons, it is difficult to extract conclusions on the human bioavailability of pomegranate husk procyanidins as these polyphenols have never been assayed for bioavailaibility in humans in any food source. It is also noteworthy that among the polyphenols present in pomegranates, procyanidins (condensed tannins) constitute far lower quantities than ellagitannins (hydrolysable tannins).

3.5 CONCLUSIONS

The biological activity and physiological effects of pomegranate polyphenols have to be consistent with their bioavailability and metabolism. Thus conclusions drawn from *in vitro* studies using cell lines that are not in direct contact with pomegranate juice polyphenols under physiological conditions should be taken with caution. This is the case of studies testing activity of punicalagin on liver or breast cancer cell lines. In these tissues, the activity of the metabolites that actually circulate in plasma (urolithins and glucuronic acid conjugates), and at the concentrations reached in plasma (mean values about 1 to 5 μM) and the tissues (detectable but generally below the quantification limit), should be evaluated. The original ellagitannins (punicalagin) and ellagic acid are much more relevant in terms of activity in the gastrointestinal tract, where they are present at significant concentrations. *In vitro* and *in vivo* studies show the relevance of these polyphenols in the control of several cancer types as chemopreventive agents.

The metabolic transformation of ellagitannins into urolithins A and B (Figure 3.4) is quite a widespread phenomenon in nature. These urolithin metabolites have previously been reported to be present in the faeces of the squirrel *Trogopterus xanthippes* and their hyaluronidase inhibitory activity was also demonstrated.[39] These metabolites have also been isolated from kidney stones in cattle suffering the "clover stone" disease that is most likely associated with a large and chronic intake of a clover species *(Trifolium subterraneum)*[40] that is probably rich in ellagitannins. Urolithins A and B have also been isolated from beaver excretions and as such are constituents of *Castoreum*.[41] *Castoreum* is a urine-based fluid from castor sacs and (or) anal-gland secretion.[42]

Apart from the hyaluronidase inhibitory activity,[39] no other potential biological activity has been attributed to the microbial metabolites, urolithins. However, it is anticipated that gut microflora metabolites urolithins are potential endocrine-disrupting molecules that could resemble other described "enterophytoestrogens" (microflora-derived metabolites with estrogenic/antiestrogenic activity) with potential estrogenic/antiestrogenic effect in the range of that reported for other well-known estrogenic compounds such as enterolactone, resveratrol, genistein, and daidzein.[43] Further research is warranted to evaluate the possible role of ellagitannins and ellagic acid as dietary "pro-phytoestrogens."

ACKNOWLEDGMENT

The authors are grateful to the Spanish MEC and Fundación Séneca for financial support of this work (AGL2003-02195, AGL2004-03989, and PB/18/FS/02).

REFERENCES

1. Gil, M.I. et al., Antioxidant activity of pomegranate juice and its relationship with phenolic composition and processing, *J. Agric. Food Chem.*, 48, 4581, 2000.
2. Plumb, G.W., et al., Antioxidant properties of gallocatechin and prodelphinidins from pomegranate peel, *Redox Report*, 7, 41, 2002.
3. Gil, M.I. et al., Changes in pomegranate juice pigmentation during ripening, *J. Sci. Food Agric.*, 68, 77, 1995.
4. Aviram, M. et al., Pomegranate juice consumption reduces oxidative stress, atherogenic modifications to plasma LDL, and platelet aggregation: studies in humans and atherosclerotic apolipoprotein E-deficient mice, *Am. J. Clin. Nutr.*, 71, 1062, 2000.
5. Filippich, L.J., Zhu, J., and Alsalamani, M.T., Hepatotoxic and nephrotoxic principles in *Terminalia oblongata, Res. Vet. Sci.*, 50, 170, 1991.
6. Teel, R.W. and Martin, R.W., Disposition of the plant phenol ellagic acid in the mouse following oral administration by gavage, *Xenobiotica*, 18, 397, 1988.
7. Doyle, B. and Griffiths, L.A., The metabolism of ellagic acid in the rat, *Xenobiotica*, 10, 247, 1980.

8. Smart, R.C. et al., Disposition of the naturally occurring antimutagenic plant phenol, ellagic acid, and its synthetic derivatives, 3-O-decylellagic acid and 3,3'-di-O-methyl-ellagic acid in mice, *Carcinogenesis,* 7, 1663, 1986.

9. Van Tassel, R., Bezoars, *Janus,* 60, 241, 1973.

10. Whitley, A.C. et al., Intestinal epithelial cell accumulation of the cancer preventive polyphenol ellagic acid — extensive binding to protein and DNA, *Biochem. Pharmacol.,* 66, 907, 2003.

11. Cerdá, B. et al., Evaluation of the bioavailability and metabolism in the rat of punicalagin, an antioxidant polyphenols from pomegranate juice, *Eur. J. Nutr.,* 42, 18, 2003.

12. Manach, C. et al., Bioavailability and bioefficacy of phytonutrients in humans. I. Review of 97 bioavailability studies, *Am. J. Clin. Nutr.,* 81, 230S, 2005.

13. Cerdá, B. et al., Repeated oral administration of high doses of the pomegranate ellagitannin punicalagin to rats for 37 days is not toxic, *J. Agric. Food Chem.,* 51, 3493, 2003.

14. Larrosa, M., Tomás-Barberán, F.A., and Espín, J.C., The dietary hydrolysable tannin punicalagin releases ellagic acid that induces apoptosis in human colon adenocarcinoma Caco-2 cells by using the mitochondrial pathway, *J. Nut. Biochem.,* 2005, in press.

15. Boukharta, M., Jalbert, G., and Castonguay, A., Biodistribution of ellagic acid and dose-related inhibition of lung tumorigenesis in A/J mice, *Nutr. Cancer.,* 18, 181, 1992.

16. Lei, F. et al., Pharmacokinetic study of ellagic acid in rat after oral administration of pomegranate leaf extract, *J. Chromatogr. B,* 796, 189, 2003.

17. McGhie, T. et al., Anthocyanin glycosides from berry fruit are absorbed and excreted unmetabolized by both humans and rats, *J. Agric. Food Chem.,* 51, 4539, 2003.

18. Wu, X., Pittman, H.E. 3rd, and Prior, R.L., Pelargonidin is absorbed and metabolized differently than cyanidin after marionberry consumption in pigs, *J. Nutr.,* 134, 2603, 2004.

19. Matuschek, M.C. et al., The jejunum is the main site of absorption for anthocyanins in mice, *J. Nutr. Biochem.,* 17, 31, 2006.

20. Talavera, S. et al., Anthocyanin metabolism in rats and their distribution to digestive area, kidney, and brain, *J. Agric Food Chem.,* 18, 3902, 2005.

21. Andres-Lacueva, C. et al., Anthocyanins in aged blueberry-fed rats are found centrally and may enhance memory, *Nutr Neurosci.,* 8, 111, 2005.

22. Tsang, C. et al., The absorption, metabolism and excretion of flavan-3-ols and procyanidins following the ingestion of a grape seed extract by rats, *Br. J. Nutr.,* 94, 170, 2005.

23. Seeram, N.P., Lee, R., and Heber, D., Bioavailability of ellagic acid in human plasma after consumption of ellagitannins from pomegranate (*Punica granatum* L.) juice, *Clin. Chim. Acta,* 348, 63, 2004.

24. Cerdá, B. et al., The potent *in vitro* antioxidant elagitannins from pomegranate juice are metabolised into bioavailable but poor antioxidant hydroxyl-6H-dibenzopyran-6-one derivatives by the colonic microflora of healthy humans, *Eur. J. Nutr.,* 43, 205, 2004.

25. Stoner, G. et al., Pharmacokinetics of anthocyanins and ellagic acid in healthy volunteers fed freeze-dried black raspberries daily for 7 days, *J. Clin. Pharmacol.,* 45, 1153, 2005.

26. Cerdá, B., Tomás-Barberán, F.A., and Espín, J.C., Metabolism of antioxidant and chemopreventive ellagitannins from strawberries, raspberries, walnuts, and oak-aged wine in humans: Identification of biomarkers and individual variability, *J. Agric. Food Chem.,* 53, 227, 2005.

27. Cerdá, B. et al., Identification of Urolithin A as a metabolite produced by human colon microflora from ellagic acid and related compounds, *J. Agric. Food Chem.*, 53, 5571, 2005.

28. Cerdá, B. et al., Pomegranate juice supplementation in chronic obstructive pulmonary disease (COPD): a 5-week randomised, double blind, placebo-controlled trial, *Eur. J. Clin. Nutr.*, 60, 245, 2006.

29. Lapidot, T. et al., Bioavailability of red wine anthocyanins as detected in human urine, *J. Agric. Food Chem.*, 46, 4297, 1998.

30. Felgines, C. et al., Strawberry anthocyanins are recovered in urine as glucuro- and sulfoconjugates in humans, *J. Nutr.*, 133, 1296, 2003.

31. Talavera, S. et al., Anthocyanins are efficiently absorbed from the stomach in anesthetized rats, *J. Nutr.*, 133, 4178, 2003.

32. Passamonti, S. et al., The stomach as a site for anthocyanins absorption from food, *FEBS Lett.*, 544, 210, 2003.

33. Wu, X., Cao, G., and Prior, R.L., Absorption and metabolism of anthocyanins in elderly women after consumption of elderberry or blueberry, *J. Nutr.*, 132, 1865, 2002.

34. Milbury, P.E., Cao, G., Prior, R.L., and Blumberg, J., Bioavailability of elderberry anthocyanins, *Mech. Ageing Develop.*, 123, 997, 2002.

35. Williamson, G. and Manach, C., Bioavailability and bioefficacy of polyphenols in humans. II. Review of 93 intervention studies, *Am. J. Clin. Nutr.*, 81, 243S, 2005.

36. Sano, A. et al., Procyanidin B1 is detected in human serum after intake of proanthocyanidin-rich grape seed extract, *Biosci. Biotechnol. Biochem.*, 67, 1140, 2003.

37. Holt, R.R. et al., Procyanidin dimer B2 [epicatechin-(4β-8)-epicatechin] in human plasma after the consumption of a flavanol-rich cocoa, *Am. J. Clin. Nutr.*, 76, 798, 2002.

38. Donovan, J.L. et al., Procyanidins are not bioavailable in rats fed a single meal containing a grapeseed extract or the procyanidin dimer B-3, *Br. J. Nutr.*, 87, 299, 2002.

39. Jeong, S.J. et al., Hyaluronidase inhibitory active 6H-dibenzo[b,d]pyran-6-ones from the faeces of *Trogopterus xanthippes*, *Planta Med.*, 66, 76, 2000.

40. Pope, G.S., Isolation of two benzocoumarins from 'clover stone', a type of renal calculus found in sheep, *Biochem. J.*, 93, 474, 1964.

41. Lederer, E., Chemistry and biochemistry of some mammalian secretions and excretions, *J. Chem. Soc.*, 2115, 1949.

42. Rosell, F., Johansen, G., and Parker, H., Eurasian beavers (Castor fiber) behavioural response to simulated territorial intruders, *Can J. Zool.*, 78, 931, 2000.

43. Larrosa, M. et al., Urolithins, ellagic acid-derived metabolites produced by human colonic microflora, exhibit estrogenic and antiestrogenic activities, *J. Agric. Food Chem.*, 54, 1611, 2006.

44. Seeram, N.P., Henning, S.M., Li, Z., Lee, R., Niu, M., Suchard, M., Schueller, H.S., and Heber, D. Pharmcokinetics and *ex vivo* bioactivities of pomegranate polyphenols in human subjects. *FASEB J.*, 20(5):A655.1, 2006.

Section 2

Health Effects

4 Protection against Cardiovascular Diseases

Bianca Fuhrman and Michael Aviram

CONTENTS

4.1 RISK FACTORS FOR ATHEROSCLEROSIS

Atherosclerosis is the major cause of morbidity and mortality in the Western world, and its pathogenesis involves complicated interactions among cells of the arterial wall, blood cells, and plasma lipoproteins.[1,2] The development of the atherosclerotic lesion involves a series of specific morphological, cellular, and molecular responses[3] that can be considered inflammatory.[4,5] Following endothelial injury, blood monocytes attach to the subendothelium and penetrate into the vessel wall, where they differentiate into macrophages, forming foam cells. Blood platelets at the site of the vascular injury, monocyte-derived macrophages, endothelial cells, and smooth muscle cells release mitogenic factors, which stimulate smooth muscle cell proliferation and migration. Smooth muscle cell proliferation in turn, together with the organization of thrombus and extracellular matrix synthesis, lead to the development of the atheromatous plaques. Macrophages, by releasing proteases such as collagenase and elastase, form an abscess in the plaque, which is covered by a thin fibrous cap. When this cap ruptures, a local thrombus is formed, and the degree and duration of thrombus formation, as well as the degree of collateral development, determines the fate of this thrombus.

Atherosclerosis is the result of multiple interactive risk factors, including hypertension, activation of blood platelets, increased plasma LDL-cholesterol level, and LDL oxidative modifications.[6,7,8]

4.1.1 HYPERTENSION

Hypertension is a major risk factor for the development of atherosclerosis. In hypertensive patients with elevated plasma renin activity, increased incidence of myocardial infarction by fivefold has been demonstrated.[9] In hypertensive patients, serum concentrations of angiotensin II (Ang-II), an active vasoconstrictor produced by the renin-angiotensin-aldosterone system (RAS), are elevated.[10] Ang-II was implicated in acceleration of atherosclerosis[11,12] not only by causing hypertension, but also by stimulating proliferation of smooth muscle cells,[13] activation of blood platelets,[14] and accumulation of cholesterol in arterial macrophages,[15–17] and by increasing the formation of reactive oxygen species (ROS), such as hydrogen peroxides and other free radicals in plasma[18,19] and in macrophages.[20] In addition, Ang-II was shown to modify LDL to yield an atherogenic lipoprotein, which is taken up by macrophages at enhanced rate through the scavenger receptors. Ang-II also enhances the uptake of oxidized LDL (Ox-LDL) by macrophages via a proteoglycan-mediated pathway.[21] In apolipoprotein E-deficient (E^0) mice it was shown that Ang-II type I receptor antagonists and inhibitors of the angiotensin converting enzyme (ACE), which inhibit the production of Ang-II from Ang-I, attenuate the enhanced development and progression of atherosclerosis.[22,23] Some of the biological activities of Ang-II were shown to be mediated by aldosterone.[24] Experimental evidence suggests that aldosterone amplifies tissue ACE activity and angiotensin receptor type 1 synthesis.[25] Moreover, aldosterone plays an important role in the pathophysiology of heart failure.[26] Clinical studies have demonstrated that ACE inhibitors significantly reduce the morbidity and mortality of patients with myocardial infarction or heart failure and reduce the incidence of recurrent myocardial infarction and ischemic events in patients with coronary artery disease, even in the absence of blood pressure lowering.[27,28] Data from a variety of animal experiments indicate that ACE inhibitors can attenuate the development of atherosclerosis by mechanisms that may be mediated through the prevention of the effect of angiotensin II and also by potentiation of bradykinin formation.[29–32]

4.1.2 PLATELET AGGREGATION

Thrombus formation is important in cardiovascular diseases.[33,34] Platelets are blood cell fragments that originate from the cytoplasm of megakaryocytes in the bone marrow and circulate in blood. They play a major role in the hemostatic process and in thrombus formation after an endothelial injury. Circulating human platelets also play an important role in the development of atherosclerosis, and increased platelet aggregation is associated with enhanced atherogenicity.[35] Recent studies have provided insight into platelet functions in inflammation and atherosclerosis.[36] Platelets release several growth factors and bioactive agents that play a central role in the development of thrombus and intimal thickening.[37] A range of molecules, present on the platelet surface or stored in platelet granules, contribute to the cross-talk among platelets and other inflammatory cells during vascular inflammation, which is involved in the development and progression of atherosclerosis.[38,39] Platelet activation can be seen indeed along the different phases of atherosclerosis. Most risk factors for atherosclerosis, including hypertension,[40] cigarette smoking,[41] diabetes,[42] and hypercholesterolemia[43,44] are able to increase the number of activated platelets in the circulation. An imbalance of the hemostatic system and persistent *in vivo* platelet activation can be observed in hypercholesterolemia and may have pathophysiological implications in the development and progression of the atherosclerotic plaques. Platelet activation influences the development of atherosclerosis, and inhibition of platelet thromboxane A_2 (TX A_2) production by aspirin[45] indomethacin,[46] or TX receptors by antagonist,[47] dramatically diminish the formation of atherosclerotic lesions in LDL receptor and apolipoprotein E knockout mice. On the other hand, in the early phase of atherosclerosis, platelet activation may be attributed to reactive oxygen species (ROS) such as superoxide, hydroxyl radical, and peroxynitrite, which are generated by some risk factors for atherosclerosis. Antioxidants decrease platelet aggregation.[48–50] Vitamin E[51–53] and polyphenols or polyphenol-rich nutrients were shown indeed to decrease platelet function *in vitro* and also *ex vivo* after supplementation to humans.[54–59]

4.1.3 BLOOD LIPIDS

The important role of high-serum cholesterol, especially a high level of low-density lipoprotein (LDL) cholesterol, as a risk factor for coronary artery disease is well established,[60,61] and lowering of serum LDL levels reduces the risk for major coronary events.[62–64] On the contrary, low levels of high-density lipoprotein (HDL) constitute an independent risk factor for cardiovascular diseases (CVD).[65–67] Epidemiological studies have shown an inverse correlation between serum HDL concentration and CVD risk.[68] HDL protects against cardiovascular disease,[69,70] and recent studies have elucidated the molecular mechanisms for its action.[71,72] HDL has been proposed to decrease CVD by stimulating reverse cholesterol transport (RCT), a process by which HDL carries excess cholesterol from peripheral cells, including foam cells in the coronary artery, back to the liver for its removal from the body.[73] Beyond reverse cholesterol transport, HDL was also shown to protect LDL from oxidation.[72] HDL inhibits the oxidation of LDL by transition metal ions, but also prevents 12-lipoxygenase–mediated formation of lipid hydroperoxides. Inhibition of LDL oxidation by HDL is usually attributed to the high content of antioxidants in this lipoprotein, to the antioxidative properties of apolipoprotein A-I, and to the presence of several enzymes, such as paraoxonase 1 (PON1),[74–76] which protect LDL from oxidation by degradation of oxidized bioactive products.

Several lines of evidence support the concept that raising HDL may provide substantial atheroprotective benefit. A rise in serum HDL by 1 mg/dL has been proposed to reduce the risk of CVD by 2 to 3%. Current approaches to increase HDL levels in plasma include dietary intervention.[77–79] Several studies indicate that light to moderate alcohol consumption, including red wine, is associated with an increase in serum HDL levels,[80,81] along with a low prevalence of coronary heart disease.

4.1.4 OXIDATIVE STRESS

The "oxidative modification of lipoproteins" hypothesis of atherosclerosis proposes that LDL oxidation plays a pivotal role in early atherogenesis.[82–92] This hypothesis is supported by evidence that LDL oxidation indeed occurs *in vivo*[93] and contributes to the clinical manifestation of atherosclerosis. The uptake of Ox-LDL via scavenger receptors promotes cholesterol accumulation and foam cell formation,[94,95] the hallmark of early atherosclerosis. In addition, Ox-LDL atherogenicity is related to recruitment of monocytes to the intima,[96] stimulation of monocyte adhesion to the endothelium,[97] and its cytotoxicity toward arterial cells.[98,99] The process of LDL oxidation appears to occur within the artery wall, and all major cells of the artery wall, including endothelial cells, smooth muscle cells and monocyte-derived macrophages can oxidize LDL.[100,101] Macrophage-mediated oxidation of LDL is a key event during early atherogenesis[102] and requires the binding of LDL to the macrophage LDL receptor.[103] The interaction of LDL with macrophages under oxidative stress activates cellular oxygenases, which can then produce ROS capable of oxidizing LDL.[104,105] Under oxidative stress, not only are LDL and the other plasma lipoproteins oxidized, but cell lipid peroxidation takes place, including arterial macrophages.[106] Such "oxidized macrophages" can easily oxidize LDL.[107] The oxidation rate of LDL was shown to be reduced by dietary antioxidant intervention.[108] The oxidative hypothesis of atherosclerosis has stimulated extensive investigation on the role of antioxidants as a possible preventive treatment for atherosclerosis. Macrophage foam cell formation during early atherogenesis is determined by the balance between pro-oxidants and antioxidants in arterial wall cells, as well as in plasma lipoproteins.[109] Epidemiological studies have demonstrated an association between increased intake of antioxidant vitamins and reduced morbidity and mortality from coronary artery disease (CAD).[110,111] The beneficial health effects attributed to the consumption of fruits and vegetables are related, at least in part, to their antioxidant activity.[112–115] Some antioxidants that prevent oxidative stress were shown to protect LDL from oxidation, and in parallel, to reduce the development of the atherosclerotic lesions.[116,117]

4.2 FLAVONOIDS AND CARDIOVASCULAR DISEASES

Flavonoids compose the largest and the most studied group of plant phenolics. Over 4000 different flavonoids have been identified to date. Flavonoids are grouped into anthocyanins and anthoxanthins. Anthocyanins are glycosides of anthocyanidin, and they are the most important group of water-soluble plant pigments, responsible for the red, blue, and purple colors of flowers and fruits. Anthoxanthins are colorless or colored white-to-yellow, and include flavonols, flavanols, flavones, flavans, and isoflavones. Flavonoids are powerful antioxidants, and their activity is related to their chemical structure.[118,119] Plant flavonoids can act as potent inhibitors of LDL oxidation,[120,121] or of macrophage oxidation.[122] Dietary consumption of flavonoids was shown to be inversely related to morbidity and mortality from coronary heart disease.[123] Moreover, an inverse association between flavonoid intake and subsequent occurrence of ischemic heart disease, or cerebrovascular disease, was shown.[124,125] Reduced morbidity and mortality from cardiovascular diseases, in spite of high intake of saturated fat among French, the so-called "French paradox,"[126] has been attributed to the regular intake of red wine in the diet. Dietary consumption of flavonoid-rich nutrients, as well as pure flavonoids, was shown to attenuate the progression of atherosclerosis in animals.[127] Reduced development of atherosclerotic lesion areas in the atherosclerotic E^0 mice was demonstrated following consumption of red wine,[128,129] licorice root extract,[130,131] grape powder,[132] or ginger extract.[133]

4.3 POMEGRANATE JUICE (PJ) INHIBITS ATHEROSCLEROTIC LESION DEVELOPMENT

The pomegranate tree, which is said to have flourished in the Garden of Eden, has been extensively used as a folk medicine in many cultures.[134,135] Edible parts of pomegranate fruits (about 50% of total fruit weight) comprise 80% juice and 20% seeds. Fresh juice contains 85% moisture, 10% total sugars, 1.5% pectin, ascorbic acid, and polyphenols.

Content of soluble polyphenols in pomegranate juice (PJ) varies within the limits of 0.2 to 1.0%, depending on the variety, and includes mainly anthocyanins (such as cyanidin-3-glycoside, cyanidin-3, 3-diglycoside, and delphindin-3- glucoside) and anthoxanthins (such as catechins, ellagic tannins, and gallic and ellagic acids).[136,137] Ellagic acid and hydrolysable ellagitannins are both implicated in protection against atherogenesis, along with their potent antioxidant capacity. Punicalagin is the major ellagitannin in PJ, and this compound is responsible for the high antioxidant activity of this juice. As a major source for polyphenolics, PJ was shown to be a very potent antioxidant against LDL oxidation and was additionally shown to inhibit atherosclerosis development in mice and in humans.[138–140] *In vivo* studies were conducted first in order to evaluate whether the active antioxidant components of PJ are absorbed. Recent studies examined the bioavailability and metabolism of punicalagin in the rat as an animal model.[141,142] Two groups of rats were studied. One group was fed with standard rat diet (n = 5) and the second one with the same diet plus 6% punicalagin (n = 5). The daily intake of punicalagin ranged from 0.6 to 1.2 g. In plasma, glucuronides of methyl ether derivatives of ellagic acid and punicalagin were detected.

Also observed in the plasma were 6H-dibenzo[b,d]pyran-6-one derivatives, especially during the last few weeks of the study. In urine, the main metabolites observed were the 6H-dibenzo[b,d]pyran-6-one derivatives, and were present as aglycones or as glucuronides. It was concluded that since only 3 to 6% of the ingested punicalagin was detected as such or as metabolites in urine and feces, the majority of this ellagitannin has to be converted to undetectable metabolites or accumulated in nonanalyzed tissues. Only traces of punicalagin metabolites were detected in liver or kidney. In humans, consuming PJ (180 mL) containing 25 mg of ellagic acid and 318 mg of hydrolysable ellagitannins (as punicalagin), ellagic acid was detected in human plasma at a maximum concentration of 32 ng/mL 1 hour postingestion, and it was completely eliminated by 4 hours.[143] Thus, active components of PJ are indeed absorbed, and subsequently affect biological processes that are related to atherogenesis protection. Upon analyzing the influence of the physiological conditions in the stomach and small intestine on the bioavailability of pomegranate bioactive compounds using an *in vitro* availability method, it was demonstrated that pomegranate phenolic compounds are available during digestion in a high amount (29%). Nevertheless, due to pH, anthocyanins are largely transformed into nonred forms or degraded.[144]

4.3.1 STUDIES IN ATHEROSCLEROTIC MICE

PJ supplementation to the atherosclerotic E^0 mice reduced the size of their atherosclerotic lesion and the number of foam cells in their lesion,[145] in comparison to control placebo-treated E^0 mice that were supplemented with water (Figure 4.1A to C). We also analyzed the therapeutic potency of PJ by its administration to E^0 mice with already-advanced atherosclerosis. Atherosclerotic E^0 mice at 4 months of age were supplemented for 2 months with 31 µL of PJ (equivalent to 0.875 µmoles of total polyphenols/mouse/day, which is equivalent to about one glass or 8 oz/human/day), and were compared to age-matched placebo-treated mice, as well as to the second control group of 4-month-old mice. Although the atherosclerotic lesion area in the PJ-treated mice was increased in comparison to the lesion size observed in the control younger mice (4 months of age), PJ supplementation was still able to reduce the mice atherosclerotic lesion size by 17%, in comparison to atherosclerotic lesions of the age-matched placebo-treated mice.[146] These results were further confirmed by de Nigris et al.,[147] who demonstrated that oral administration of PJ to hypercholesterolemic LDL-receptor-deficient mice at various stages of the disease reduced significantly the progression of atherosclerosis. Thus, PJ exhibits preventive as well as therapeutic effects against atherosclerosis.

4.3.2 STUDIES IN HUMANS

We next investigated the effects of PJ consumption by patients with carotid artery stenosis (CAS) on carotid lesion size, in association with changes in oxidative stress.[148] Ten patients were supplemented with PJ for up to one year, and nine CAS

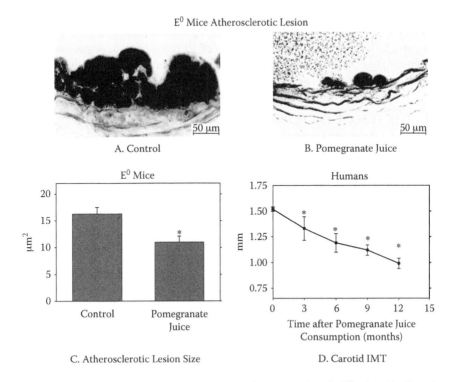

C. Atherosclerotic Lesion Size   D. Carotid IMT

FIGURE 4.1 Pomegranate juice consumption by the atherosclerotic E^0 mice (A–C) or by patients with carotid artery stenosis (D) reduces atherosclerotic lesion size and carotid IMT. Thirty E^0 mice and 10 patients with severe CAS were supplemented with PJ concentrate (12.5 μL/mouse/day and 50 mL/day, respectively) for 9 weeks or for 1 year, respectively. Photomicrographs of typical foam cells from E^0 mice that consumed water (control, A) or PJ (B) are presented. Mean (± SEM) effect of PJ consumption on mice atherosclerotic lesion area (C) and on human common carotid artery IMT (D) are shown. *$p < 0.01$ (after vs. before PJ consumption).

patients who did not consume PJ served as a control group. Blood samples were collected before treatment and after 3, 6, 9, and 12 months of PJ consumption. Patients' carotid intima-media thickness (IMT) was compared between the PJ group and the control group. While in the control group (no PJ) IMT increased by 10% after 1 year, PJ consumption resulted in a significant IMT reduction, by up to 43% (Figure 4.1D). Analysis of carotid IMT before and during PJ consumption revealed a gradual reduction of 20%, 27%, 28%, and 38% in the left carotid artery, and a reduction of 6%, 25%, 28%, and 32% in the right carotid IMT, as observed after 3, 6, 9, and 12 months of PJ consumption, respectively, in comparison to "0 time." On examination of the internal carotid arteries, flow velocities were calculated at the stenotic sites, and expressed by peak systolic velocity (PSV) and end diastolic velocity (EDV). The ultrasound outcome data were the change over time in maximal IMT, which was measured in the same preselected carotid artery segments. Twelve months of PJ consumption resulted in PSV reduction by 12% and 28% in the left and right carotid arteries, respectively. Carotid EDV decreased in the left carotid artery by 11%, 14%, 17%, and 33%, and in the right carotid artery by 20%, 26%, 40%, and 52% after 3, 6, 9, and 12 months of PJ consumption, respectively. Thus, PJ consumption by patients with CAS decreases atherosclerotic lesion size, and these effects could be related to the potent antioxidant characteristics of PJ.

4.4 THE EFFECT OF POMEGRANATE JUICE ON HYPERTENSION

Some antioxidants such as vitamin C, vitamin E, β-carotene and coenzyme Q were recently shown to possess hypotensive properties.[149–152] As PJ possesses very potent antioxidative properties, we questioned the effect of PJ on blood pressure and on serum ACE activity. PJ consumption by CAS patients significantly ($p < 0.05$) reduced their systolic blood pressure after 1 year by 18% (from 174 ± 22 to 143 ± 17 mmHg)[148] (Figure 4.2A). In contrast, PJ consumption had no significant effect on the patient's diastolic blood pressure (81 ± 3 before treatment vs. 81 ± 2 mmHg after one year of PJ consumption). Seven hypertensive males and three females with mean blood pressure levels of $155 \pm 7/83 \pm 7$ mmHg were also studied.[153] The patients were administered PJ (50 mL concentrate containing 1.5 mmol of total polyphenols per day) for a period of 2 weeks. PJ consumption resulted in a small (5%) but significant ($p < 0.05$) reduction in the systolic blood pressure (Figure 4.2B). In seven out of ten hypertensive patients studied, serum ACE activity was significantly decreased by 36% after 2 weeks of PJ consumption (Figure 4.2C). To assess a possible direct effect of PJ on serum ACE activity, increasing concentrations of PJ were added to human serum and incubated for 15 min at 37°C. A PJ dose-dependent inhibitory effect, up to 31%, on serum ACE activity was obtained (Figure 4.2D). This effect may be secondary to the ability of PJ-associated antioxidants, such as complexed tannins, to inhibit ACE activity. Thus, PJ possesses a direct inhibitory effect on serum ACE activity. Because ACE inhibitors are metabolized by cytochrome P-450 enzymes, serum ACE activity can be significantly affected by modulation of P-450 enzyme activity.[154] Therefore, we next analyzed the effect of PJ on cytochrome P-450 enzymes.[155] PJ decreased the activities of cytochrome P-450 3A4, 2D6, 2E1, and 2B6 by 40%, 30%, 20%, and 60%, respectively. In hypertensive patients treated with ACE inhibitors, the inhibitory effect of PJ consumption on the P-450 enzymes can possibly decrease P-450-mediated drug (the ACE inhibitor) breakdown and hence, serum ACE activity may be further decreased in these treated patients. We have indeed observed in three hypertensive patients treated with the ACE inhibitor fosinopril (20 mg/day for 1 month) that their serum ACE activity decreased by 26% after 2 weeks of PJ consumption (50 mL of concentrate containing 1.5 mmoles of total polyphenols/day), in comparison to the patients treated with fosinopril but who did not receive PJ. Taken together, the results on the inhibitory effect of PJ on serum ACE activity on the one hand, and on cytochrome P-450 enzymes on the other hand, suggest that PJ may affect ACE activity also indirectly, secondary to its inhibitory effect on the cytochrome P-450 enzymes. As ACE activity

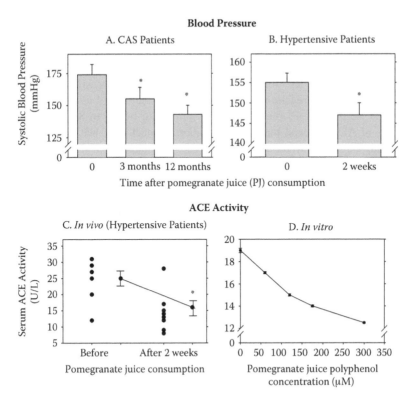

FIGURE 4.2 The effect of pomegranate juice on blood pressure and ACE activity. Patients with carotid artery stenosis (A) or patients with hypertension (B) were administered PJ for 12 months and for 2 weeks, respectively. Blood pressure was measured before PJ consumption (0) and after 3 and 12 months, or 2 weeks, respectively. ACE activity was measured in serum derived from hypertensive patients before and after PJ consumption for 2 weeks (C), and *in vitro* in normal serum that was preincubated with increasing concentrations of PJ (D). Results are mean ± SEM. *$p < 0.01$ (after vs. before PJ consumption).

is related to enhanced blood pressure and accelerated atherosclerosis, our data may suggest an additional important antiatherogenic property to PJ consumption.

4.5 THE EFFECT OF POMEGRANATE JUICE ON PLATELET ACTIVITY

Circulating human platelets play an important role in the development of atherosclerosis, and increased platelet aggregation is associated with enhanced atherogenicity. Platelet activation was shown to be associated with oxidative stress. To study whether PJ can inhibit platelet aggregation, we supplemented 13 healthy, nonsmoking men with 50 mL/day of PJ concentrate (contains 1.5 mmoles of total polyphenols) for a period of 2 weeks.[145] Following 2 weeks of PJ consumption, a significant

($p < 0.02$) 11% reduction in collagen-induced platelet aggregation was noted, in comparison to platelet aggregation prior to PJ consumption at the beginning of the study. The ability of PJ consumption to reduce platelet aggregation in humans was supported by a direct effect of PJ on platelet aggregation *in vitro*. We incubated platelet-rich plasma (PRP) for 30 minutes at 37°C with increasing concentrations of PJ, after which aggregation was induced by the addition of collagen (2 µg/mL) or of ADP (2 µM). A PJ dose-dependent inhibition, by up to 87% and 94%, of collagen- or ADP-induced platelet aggregation, respectively, was observed. These effects may be related to an interaction of PJ constituents with the platelet surface binding sites for collagen or ADP. It may also be that the antioxidative properties of PJ constituents, as demonstrated by their ability to scavenge free radicals, can attenuate oxidative stress-induced platelet activation. In a recent study it was demonstrated that PJ consumption by 28 fasted, healthy adult subjects (6 to 9 mL/kg) inhibited platelet function, as measured by a prolongation of epinephrine/collagen-induced clotting time.[156] Furthermore, chronic exposure of cultured human aortic endothelial cells (HAEC) to PJ increased prostacyclin synthesis by 61%, further evidencing that PJ can reduce atherogenicity by inhibiting platelet activation.

4.6 THE EFFECT OF POMEGRANATE JUICE ON SERUM LIPID PATTERN

Pomegranate juice concentrate (50 mL/day, equivalent to 1.5 mmol of total polyphenols) administration to healthy men for a period of 2 weeks had no significant effect on plasma lipid profile, including total cholesterol, LDL-cholesterol, VLDL-cholesterol, HDL-cholesterol, and triglyceride concentrations. Similarly, no significant effect could be demonstrated on plasma lipid concentrations in the E° mice after consumption of PJ, in comparison to control mice that consumed only water.[145] The effect of PJ consumption on serum lipid profile was also studied in diabetic patients with hyperlipidemia.[157] In this study, 22 type II diabetic patients with cholesterol levels higher than 5.2 mmol/L and triglyceride levels higher than 2.3 mmol/L consumed 40 g/day of concentrated PJ for 8 weeks. After consumption of concentrated PJ, significant reductions were seen in total cholesterol, in LDL cholesterol, and in the ratio of LDL-cholesterol/HDL-cholesterol, as well as total cholesterol/HDL-cholesterol. No significant changes were measured in serum triglyceride or HDL-cholesterol levels. However, a similar study conducted in our laboratory with diabetic patients did not show significant effects of PJ consumption on serum lipid levels (unpublished data).

4.7 THE EFFECT OF POMEGRANATE JUICE ON OXIDATIVE STRESS IN SERUM

4.7.1 SERUM LIPID PEROXIDATION

Human plasma obtained from healthy men after 2 weeks of PJ consumption (50 mL PJ concentrate/day, equivalent to 1.5 mmol total polyphenols) demonstrated a small but significant ($p < 0.01$) 16% decrease in susceptibility to free-radical-induced lipid peroxidation, in comparison to plasma obtained prior to PJ consumption, as measured by lipid peroxides formation (Figure 4.3A), or as total antioxidant status (TAS) in serum (Figure 4.3B). To determine the effect of increasing or decreasing the dosages of PJ on plasma lipid peroxidation and to analyze PJ capability to maintain its effect after termination of juice consumption, three subjects were further studied. Supplementation of 20 ml of PJ concentrate/day for 1 week resulted in a significant decrease of 11% in plasma lipid peroxidation, compared to plasma obtained prior to PJ consumption. Supplementation of 50 mL PJ concentrate/day for 1 more week exhibited a further 21% decrease in plasma lipid peroxidation. However, a further increase in the supplemented PJ to 80 mL of PJ concentrate/day for an additional 1 week did not further inhibit plasma susceptibility to lipid peroxidation. Gradual decreasing of the PJ dosage in these three subjects down to 40 mL/day for one week, and then to 20 mL/day for additional 2 weeks, did not significantly affect plasma lipid peroxidation, which remained low in comparison to the levels obtained after supplementation of 80 mL of PJ concentrate/day. Two weeks after cessation of PJ supplementation, the reduced rate of plasma susceptibility to lipid peroxidation was sustained. After a further 4 weeks with no PJ consumption, plasma lipid peroxidation returned to the higher values obtained before PJ consumption.

The effect of PJ consumption by patients with CAS on their serum oxidative state was measured also as serum concentration of antibodies against Ox-LDL.[148] A significant ($p < 0.01$) reduction in the concentration of antibodies against Ox-LDL by 24% and 19% was observed after 1 and 3 months of PJ consumption, respectively (from 2070 ± 61 EU/mL before treatment to 1563 ± 69 and 1670 ± 52 EU/mL after 1 and 3 months of PJ consumption, respectively). Total antioxidant status (TAS) in serum from these patients was substantially increased by 2.3-fold (from 0.95 ± 0.12 nmol/L at baseline, up to 2.20 ± 0.25 nmol/L after 12 months of PJ consumption). These results indicate that PJ administration to patients with CAS substantially reduced their serum oxidative status, and could thus inhibit plasma lipid peroxidation. The susceptibility of the patient's plasma to free-radical-induced oxidation decreased after 12 months of PJ consumption by 62% (from 209 ± 18 at baseline to 79 ± 6 nmol of peroxides/mL).

PJ consumption exhibited antioxidative effects also when administered to E° mice.[145] The basal oxidative state, measured as lipid peroxides in plasma of control E° mice (that did not consume PJ), increased gradually during aging from 260 nmol/mL of plasma at 6 weeks of age, to 309 and 535 nmol/mL of plasma after 9 and 14 weeks of age, respectively. Following PJ consumption, plasma lipid peroxidation was markedly reduced, and this effect was PJ concentration-dependent (Figure 4.3D). Similarly, serum total antioxidant status was higher in E° mice that consumed PJ, in comparison to control mice, and this effect was again juice concentration-dependent (Figure 4.3E).[145]

Pomegranate peel extracts have also been shown to possess significant antioxidant activity. Feeding albino rats of the Wistar strain with a dried methanolic extract from pomegranate peels at 50 mg/kg (in terms of catechin equivalents) followed by carbon tetrachloride (CCl_4)-induced oxidative stress, resulted in preservation of catalase, peroxidase, and superoxide dismutase (SOD) to values comparable with

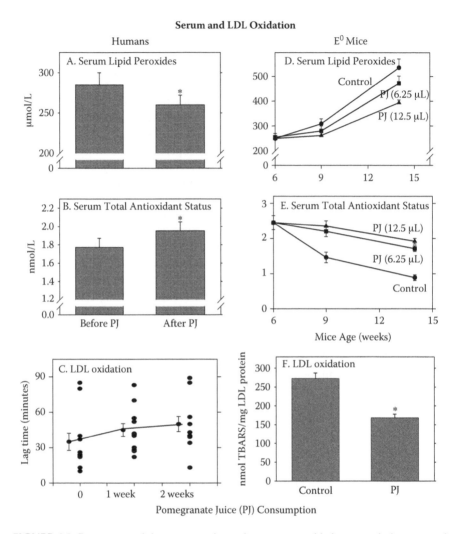

FIGURE 4.3 Pomegranate juice consumption reduces serum oxidative stress in humans and in atherosclerotic E^0 mice. Mean ($\pm$ SD) effect of 2 and 9 weeks of PJ supplementation to 13 healthy men and to E^0 mice on (A and D) the susceptibility of serum to radical-induced lipid peroxidation, (B and E) serum total antioxidant status, and (C and F) copper ion-induced LDL oxidation, is shown. *$p < 0.01$ (after vs. before PJ consumption in humans, and PJ vs. control in mice).

control values, whereas lipid peroxidation was brought back by 54% as compared to control.[158]

4.7.2 LDL AND HDL OXIDATION

Consumption of PJ for 1 and 2 weeks by healthy volunteers increased the resistance of their LDL to copper ion-induced oxidation, as shown by a prolongation of the lag time required for the initiation of LDL oxidation, by 29% and 43%, in comparison to LDL obtained prior to juice consumption (Figure 4.3C).[145] Similarly, the resistance of their HDL to copper ion-induced oxidation also gradually increased after PJ consumption, as shown by a prolongation in the lag time required for the initiation of HDL oxidation from 37 ± 2 minutes to 45 ± 6 minutes before and 2 weeks after PJ consumption, respectively. PJ consumption by patients with CAS resulted in a significant reduction in the basal level of LDL-associated lipid peroxides by 43%, 89%, 86%, and 90% after 3, 6, 9, and 12 months of PJ consumption, respectively, and in parallel it increased the resistance of LDL to copper ion-induced oxidation.[148] This was demonstrated by reduced formation of lipid peroxides in LDL during its incubation with copper ions (by 40%, 49%, 57%, and 59% after 3, 6, 9, and 12 months of PJ consumption, respectively). PJ consumption also decreased the propensity of LDL derived from E° mice to copper ion-induced oxidation.

In E° mice that consumed 6.25 µL/day or 12.5 µL/day of PJ concentrate for a period of 2 months, LDL oxidation was delayed by 100 minutes and by 120 minutes, respectively, in comparison to LDL obtained before juice administration. Determination of the extent of LDL oxidation by the TBARS assay revealed a significant inhibition after PJ consumption (Figure 4.3F). Furthermore, the progressive increase with age in the susceptibility of the mice LDL to oxidation was significantly attenuated by PJ consumption, in a dose-dependent manner.[145]

4.7.3 Paraoxonase 1 (PON1)

The increased resistance of LDL and of HDL to oxidation after PJ administration to healthy subjects or to patients with CAS could have also resulted from increased serum HDL-associated paraoxonase activity. Indeed, a significant 18% increase in serum paraoxonase (PON 1) activity was monitored in healthy subjects after PJ consumption for a period of 2 weeks.[145] In CAS patients, serum paraoxonase 1 (measured as arylesterase activity) significantly increased by 11%, 42%, 49%, and 83% after 3, 6, 9, and 12 months of PJ consumption, respectively.[148] Similar to the results in humans, a significant 43% increase in serum paraoxonase activity was also observed in E⁰ mice after PJ consumption for a period of 2 months, in comparison to serum paraoxonase activity observed in the placebo-treated mice.[146] The increase in serum paraoxonase activity may be a direct effect of PJ, as well as an effect secondary to PJ-mediated reduction in lipid peroxides. It was previously demonstrated that paraoxonase is inactivated by oxidized lipids,[159] and its activity is preserved by antioxidants, including the red wine flavonoids or the licorice-derived isoflavan glabridin. PJ contains very potent antioxidants and, unlike other nutrients,[159] it not only preserves serum PON1, but even increased the enzyme's activity.

4.8 THE EFFECT OF POMEGRANATE JUICE ON MACROPHAGE ATHEROGENICITY

Oxidative stress, which has been implicated in the pathogenesis of atherosclerosis,[2,160] has been shown to considerably attack lipids, not only in LDL but also in arterial macrophages.[106,107] We have previously shown that "lipid-peroxidized mac-rophages" exhibit atherogenic characteristics, including increased ability to oxidize LDL and to take up Ox-LDL.[161]

LDL oxidation by macrophages is considered to be the hallmark of early athero-genesis, and it is associated with cellular uptake of oxidatively modified LDL, leading to macrophage cholesterol accumulation and foam cell formation. We thus studied the effect of dietary consumption of PJ by E° mice on macrophage athero-genicity, including macrophage lipid peroxidation and subsequently macrophage activities related to foam cell formation, such as cell-mediated oxidation of LDL and cellular uptake of lipoproteins.

4.8.1 MACROPHAGE OXIDATIVE STRESS

We have demonstrated that the carotid lesion derived after endartherectomy from CAS patients who consumed PJ (Figure 4.4A), as well as mouse peritoneal mac-rophages (MPM) isolated from E° mice after consumption of PJ concentrate (12.5 μL/mouse/day, equivalent to 0.35 μmoles of total polyphenols) for a period of 2 months (Figure 4.4B), contained less lipid peroxides, in comparison to carotid lesion from patients who did not consume PJ, or to MPM from control E° mice, respectively.[145] Incubation of the human carotid lesion or of E^0 mouse peritoneal macrophages with LDL (100 μg of protein/mL) for 18 hours under oxidative stress (in the presence of copper ions) revealed that PJ consumption resulted in 43% and 82% reduced capacity of the lesion or the macrophages to oxidize LDL, respectively (Figure 4.4C and D). The mechanism responsible for this effect was associated with inhibition of the translocation to the macrophage plasma membrane of the NADPH oxidase cytosolic factor p-47, and hence, inhibition of NADPH oxidase activation. As a result of this, a 49% reduction in superoxide anion release from the macrophages and a 25% elevation in cellular glutathione content were observed. These effects could also be related to the decreased levels of macrophage-associated lipid perox-ides after PJ consumption, in comparison to macrophages isolated from control E^0 mice that consumed placebo.

On a molecular basis, PJ could interfere with macrophage oxidative status and macrophage-mediated LDL oxidation by affecting redox-sensitive gene transcription. The activation of Nuclear Factor kappa-B (NFκ-B), the oxidative stress responsive transcription factor, has been linked with a variety of inflammatory diseases, includ-ing atherosclerosis. Extensive research in the last few years, reviewed by Aggarwal and Shishoda,[162] has shown that the pathway that activates NFκ-B can be inhibited by phytochemicals, including those present in pomegranate, thus providing a bene-ficial effect against atherosclerosis development. It was demonstrated that pome-granate wine (PJ fermented with yeast and dealcoholized) inhibits oxidation of endothelial cells induced by TNF-α and acts as a potent inhibitor of NFκ-B activation in these cells.[163] Pomegranate fermented juice and pomegranate cold-pressed seed oil flavonoids were also shown to inhibit eicosanoid enzyme activity.[164] Flavonoids extracted from pomegranate cold-pressed seed oil showed 31 to 44% inhibition of sheep cyclooxygenase and 69 to 81% inhibition of soybean lipoxygenase. Flavonoids extracted from pomegranate fermented juice also showed 21 to 30% inhibition of

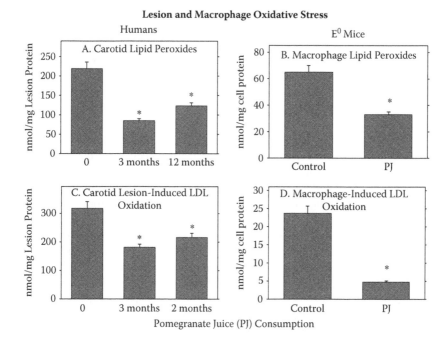

FIGURE 4.4 Pomegranate juice consumption reduces atherogenicity of human carotid lesion and of E^0 mice macrophages. Mean ($\pm$ SD) of the effect of PJ consumption on lipid peroxides in carotid lesion (A) and in mouse peritoneal macrophages (B), and LDL oxidation by carotid lesion (C) and by mouse macrophages (D). $*p < 0.01$ (after vs. before PJ consumption in humans, and PJ vs. control in mice).

soybean lipoxygenase. Recently, it was demonstrated that PJ decreased the activation of the redox-sensitive genes ELK-1 and p-JUN and increased eNOS expression in cultured endothelial cells that were exposed to shear stress, as well as in atherosclerotic-prone areas of hypercholesterolemic mice.[147]

The PJ-mediated reduction in the transcription of several key redox enzymes, including cyclooxygenase, lipoxygenase, and NO synthase could be the result of intracellular oxidation suppression. Through this mechanism, as well as via the suppression of lipoxygenase-catalyzed leukotriene formation, PJ may act also as an anti-inflammatory agent in addition to its major role as an antioxidant.

4.8.2 MACROPHAGE CHOLESTEROL METABOLISM

As mentioned before, macrophage cholesterol accumulation and foam cell formation are the hallmark of early atherogenesis. Cholesterol accumulation in macrophages can result from impaired balance between external and internal cholesterol sources. LDL, which undergoes oxidative modification, is an important external source for macrophage accumulated cholesterol.

Ox-LDL is taken up by macrophages at enhanced rate via scavenger receptors,[85,88] which, unlike the LDL receptor, are not down-regulated by intracellular cholesterol content,[165] and therefore lead to accumulation of cholesterol in the cells. Macrophage cholesterol from internal sources originates from cholesterol biosynthesis. The enzyme 3-hydroxy-3 methylglutaryl coenzyme A (HMGCoA) reductase catalyzes the rate-limiting step in cholesterol biosynthetic pathway,[166] and it is subjected to a negative feedback regulation by the cellular cholesterol content. In addition to cellular uptake of lipoproteins and to cholesterol biosynthesis, macrophage cholesterol accumulation can also result from a decreased efflux of cholesterol from the cells.[167]

Since PJ was shown to inhibit macrophage-foam cell formation and the development of atherosclerotic lesions, we analyzed the effect of PJ consumption on cellular processes that lead to macrophage cholesterol accumulation. We have demonstrated that the cellular uptake of Ox-LDL, measured as cellular lipoprotein binding, cell-association, and degradation, by MPM derived from E° mice that consumed 12.5 µL of PJ concentrate/mouse/day for a period of 2 months, was significantly reduced, by 16%, 22%, and 15%, respectively, in comparison to Ox-LDL binding, cell-association, and degradation obtained by MPM from control E° mice. Cellular cholesterol esterification rate (another atherogenic property of macrophages) in MPM isolated from PJ-treated mice was found to be 80% lower compared with age-matched, placebo-treated mice. Finally, PJ treatment significantly increased, by 39%, cholesterol efflux from macrophages compared with the cholesterol efflux rate from MPM harvested from the placebo-treated mice. Taken together, all these anti-atherogenic effects lead to reduced accumulation of cholesterol in macrophages.

In vitro studies clearly show that PJ exhibits direct antiatherogenic effects on macrophages. Preincubation of macrophages in culture (J774A.1 cell line) with PJ resulted in a significant ($p < 0.01$) reduction in Ox-LDL degradation by 40%.[168] On the contrary, PJ had no effect on macrophage degradation of native LDL, and also none on macrophage cholesterol efflux. Macrophage cholesterol biosynthesis, however, was inhibited by 50% after cell incubation with PJ. This inhibition, unlike statin action, was not mediated by an effect on HMGCoA reductase along the cholesterol biosynthetic pathway. We conclude that PJ suppresses Ox-LDL uptake by macrophages and cellular cholesterol biosynthesis, leading to attenuation in cellular cholesterol accumulation and foam cell formation.

4.9   PERSPECTIVES AND FUTURE DIRECTIONS

Our current view on the major pathways by which pomegranate polyphenols reduce macrophage foam cell formation and the development of advanced atherosclerosis, thereby protecting against heart diseases, is summarized in Figure 4.5. Pomegranate polyphenols can protect LDL against cell-mediated oxidation via two pathways, including direct interaction of the polyphenols with the lipoprotein, indirect effect through polyphenols accumulation in arterial macrophages, or both. Pomegranate polyphenols were shown to reduce the capacity of macrophages to oxidatively modify LDL, due to inhibition of LDL oxidation by scavenging reactive oxygen species (ROS) and reactive nitrogen species (RNS), and also due to polyphenol accumulation in arterial macrophages, inhibition of macrophage lipid peroxidation, and the formation of lipid peroxide-rich macrophages when polyphenols accumulate

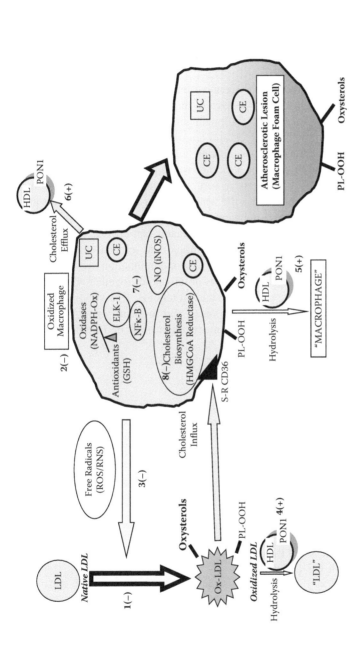

FIGURE 4.5 Major pathways by which pomegranate polyphenols inhibit macrophage foam cell formation and atherosclerosis. PJ polyphenols affect LDL directly by their interaction with the lipoprotein and inhibition of LDL oxidation (1). PJ polyphenols can also protect LDL indirectly, by their accumulation in arterial cells and protection of arterial macrophages against oxidative stress (2). This latter effect is associated with inhibition of the formation of "oxidized macrophages" and reduction in the capacity of macrophages to oxidize LDL (3). In addition, PJ polyphenols preserve or even increase paraoxonase activity, thereby increasing hydrolysis of lipid peroxides in Ox-LDL ("LDL") (4) or in oxidized macrophages in the atherosclerotic lesion ("Macrophage") (5), or increasing HDL-mediated efflux from macrophages (6), leading to attenuation in the progression of atherosclerosis. PJ polyphenols can also reduce the oxidative capacity of macrophages by reducing the activation of the redox-sensitive genes ELK-1 and NFκ-B, and increasing activation of inducible NO (iNOS) (7). Furthermore, PJ inhibits cholesterol biosynthesis in macrophages (8), thus reducing cholesterol accumulation in macrophages and their conversion into foam cells. ROS, reactive oxygen species; RNS, reactive nitrogen species; PL-OOH, phospholipid hydroperoxides; S-R, scavenger receptors; CE, cholesterol ester; UC, unesterified cholesterol; NO, nitric oxide; iNOS, inducible nitric oxide synthase; (+), stimulation; (−), inhibition.

in the arterial macrophages. Furthermore, pomegranate polyphenols increase serum paraoxonase activity, resulting in the hydrolysis of lipid peroxides in oxidized lipoproteins and in atherosclerotic lesion. Moreover, PJ has a remarkable effect on the atherogenicity of macrophages. PJ was demonstrated to reduce accumulation of cholesterol in these cells due to inhibition of cellular cholesterol biosynthesis and cellular uptake of Ox-LDL, and to reduce the oxidative capacity of the cells towards LDL.

All these antioxidative and antiatherogenic effects of pomegranate polyphenols were demonstrated *in vitro,* as well as *in vivo* in humans and in the atherosclerotic apolipoprotein E-deficient mice. Dietary supplementation of pomegranate juice rich in polyphenols to patients with severe carotid artery stenosis or with diabetes[168] or to atherosclerotic mice resulted in a significant inhibition in the development of atherosclerotic lesions, and this may be attributed to the protection against oxidation of lipids in the arterial wall as well as in serum. Furthermore, administration of pomegranate byproduct, which includes the whole pomegranate fruit left after juice preparation, to the apolipoprotein E-deficient mice reduced the atheroslerotic lesion size by up to 57%. Since a combination of antioxidants can provide a wider range of free-radical scavenging than an individual antioxidant, clinical and nutritional studies in humans should be directed toward the use of combinations of several types of dietary antioxidants, including combinations of flavonoids together with the other nutritional antioxidants, such as vitamin E and carotenoids. It is very important to use reliable biological markers of oxidative stress, and to identify populations suitable for antioxidant treatment, as antioxidant treatment may be beneficial only in subjects who are under oxidative stress.

REFERENCES

1. Glass, C.K. and Witztum, J.L., Atherosclerosis: The road ahead, *Cell,* 104, 503, 2001.
2. Aviram, M., Review of human studies on oxidative damage and antioxidant protection related to cardiovascular diseases, *Free. Radic. Res.,* 33, S85, 2000.
3. Badimon, J.J. et al., A multifactorial disease, Circulation, 87, 13, 1993.
4. Ross, R., Atherosclerosis is an inflammatory disease, *Am. Heart J.,* 138, S419, 1999.
5. Ross, R., Atherosclerosis — an inflammatory disease, *N. Engl. J. Med.,* 340, 115, 1999.
6. Homma, Y., Predictors of atherosclerosis, *J. Atheroscler. Thromb.,* 11, 265, 2004.
7. Lusis, A.J., Atherosclerosis, *Nature,* 407, 233, 2000.
8. Stocker, R. and Keaney, J.F., Jr., Role of oxidative modifications in atherosclerosis, *Physiol. Rev.,* 84, 1381, 2004.
9. Alderman, M.H. et al., Association of the renin-sodium profile with the risk of myocardial infarction in patients with hypertension, *N. Engl. J. Med.,* 324, 1098, 1991.
10. Chobanian, A.V. and Dzau, V.J., Renin angiotensin system and atherosclerotic vascular disease, in *Atherosclerosis and Coronary Artery Disease,* Fuster, V., Ross, R., and Topol, E.J., eds. Vol. 1. Philadelphia: Lippincott-Raven, 1996, 237.
11. Keidar, S., Angiotensin, LDL peroxidation and atherosclerosis, *Life Sci.,* 63, 1, 1998.

12. Daugherty, A., Manning, M.W., and Cassis, L.A., Angiotensin II promotes athero-sclerotic lesions and aneurysms in apolipoprotein E-deficient mice, *J. Clin. Invest.,* 105, 1605, 2000.

13. Weber, H., Taylor, D.S., and Molloy, C.J., Angiotensin II induces delayed mitogenesis cellular proliferation in rat aortic smooth muscle cells. Correlation with the expression of specific endogenous growth factors and reversal by suramin, *J. Clin. Invest.,* 93, 788, 1994.

14. Larsson, P.T., Schwieler, J.H., and Wallen, N.H., Platelet activation during angiotensin II infusion to healthy volunteers, *Blood Coagul. Fibrinolysis.,* 11, 61, 2000.

15. Keidar, S. et al., Angiotensin II administration to atherosclerotic mice increases macrophage uptake of oxidized LDL: a possible role for interleukin-6, *Arterioscler. Thromb. Vasc. Biol.,* 21, 1464, 2001.

16. Keidar, S., Kaplan, M., and Aviram, M., Angiotensin II modified LDL is taken up by macrophages via the scavenger receptor leading to cellular cholesterol accumula-tion, *Arterioscler. Thromb. Vasc. Biol.,* 16, 122, 1996.

17. Keidar, S. et al., Angiotensin II atherogenicity in apolipoprotein E deficient mice is associated with increased cellular cholesterol biosynthesis, *Atherosclerosis,* 146, 257, 1999.

18. Lacy, F., O'Connor, D.T., and Schmid-Schonbein, G.W., Plasma hydrogen peroxide production in hypertensive and normotensive subjects at genetic risk of hypertension, *J. Hypertens.,* 16, 291, 1998.

19. Swei, A. et al., Oxidative stress in the Dahl hypertensive rat, *Hypertension,* 30, 1628, 1997.

20. Keidar, S. et al., Angiotensin II stimulates macrophage-mediated oxidation of low density lipoprotein, *Atherosclerosis,* 115, 201, 1995.

21. Keidar, S., and Attias, J., Angiotensin II injection into mice increases the uptake of oxidized LDL by their macrophages via a proteoglycan-mediated pathway, *Biochem. Biophys. Res. Commun.,* 239, 63, 1997.

22. Keidar, S. et al., The angiotensin II receptor antagonist, losartan, inhibits LDL lipid peroxidation and atherosclerosis in apolipoprotein E deficient mice, *Biochem. Bio-phys. Res. Commun.,* 236, 622, 1997.

23. Hayek, T. et al., Antiatherosclerotic and antioxidative effects of captopril in apolipo-protein E deficient mice, *J. Cardiovasc. Pharmacol.,* 31, 540, 1998.

24. Xiao, F., Puddefoot, J.R., and Vinson, G.P., Aldosterone mediates angiotensin II stimulated rat vascular smooth muscle cell proliferation, *J. Endocrinol.,* 165, 533, 2000.

25. Ullian, M.E., Schelling, J.R., and Linas, S.L., Aldosterone enhances angiotensin II receptor binding and inositol phosphate responses, *Hypertension,* 20, 67, 1992.

26. Dzau, V.J. et al., Relation of the renin-angiotensin-aldosterone system to clinical state in congestive heart failure, *Circulation,* 63, 645, 1981.

27. Bertrand, M.E., Provision of cardiovascular protection by ACE inhibitors: a review of recent trials, *Curr. Med. Res. Opin.,* 20, 1559, 2004.

28. Pitt, B. et al., The effect of spironolactone on morbidity and mortality in patients with severe heart failure. Randomized Aldactone Evaluation Study Investigators, *N. Engl. J. Med.,* 341, 709, 1999.

29. Keidar, S. et al., Aldosterone administration to mice stimulates macrophage NADPH oxidase and increases atherosclerosis development: a possible role for angiotensin-converting enzyme and the receptors for angiotensin II and aldosterone, *Circulation,* 109, 2213, 2004.

30. Hayek, T. et al., Omapatrilat decreased macrophage oxidative status and atherosclerosis progression in atherosclerotic apolipoprotein E-deficient mice, *J. Cardiovasc. Pharmacol.*, 43, 140, 2004.

31. Hayek, T. et al., Tissue angiotensin-converting-enzyme (ACE) deficiency leads to a reduction in oxidative stress and in atherosclerosis: studies in ACE-knockout mice type 2, *Arterioscler. Thromb. Vasc. Biol.*, 23, 2090, 2003.

32. Keidar, S. et al., Effect of eplerenone, a selective aldosterone blocker, on blood pressure, serum and macrophage oxidative stress, and atherosclerosis in apolipoprotein E-deficient mice, *J. Cardiovasc. Pharmacol.*, 41, 955, 2003.

33. Marutsuka, K. et al., Role of thrombogenic factors in the development of atherosclerosis, *J. Atheroscler. Thromb.*, 12, 1, 2005.

34. Warkentin, T.E., Hemostasis and atherosclerosis, *Can. J. Cardiol.*, 11, Suppl C, 29, 1995.

35. Fuster, V. et al., Atherosclerotic plaque rupture and thrombosis. Evolving concepts, *Circulation*, 82(3 Suppl), II47, 1990.

36. Huo, Y. and Ley, K.F., Role of platelets in the development of atherosclerosis, *Trends Cardiovasc. Med.*, 14, 18, 2004.

37. Fuster, V., Mechanisms of arterial thrombosis: foundation for therapy, *Am. Heart. J.*, 135, S361, 1998.

38. Aviram, M., Platelets and the arterial wall lesion, *Curr. Opin. Lipidol.*, 3, 344, 1992.

39. Aviram, M., LDL-platelet interaction under oxidative stress induces macrophage foam cell formation, *Thromb. Haemost.*, 74, 560, 1995.

40. Nityanand, S. et al., Platelets in essential hypertension, *Thromb. Res.*, 72, 447, 1993.

41. Nowak, J. et al., Biochemical evidence of a chronic abnormality in platelet and vascular function in healthy individuals who smoke cigarettes, *Circulation*, 76, 6, 1987.

42. Manduteanu, I. et al., Increased adhesion of human diabetic platelets to cultured valvular endothelial cells, *J. Submicrosc. Cytol. Pathol.*, 24, 539, 1992.

43. Aviram, M. et al., Interactions of platelets, macrophages, and lipoproteins in hypercholesterolemia: antiatherogenic effects of HMG-CoA reductase inhibitor therapy, *J. Cardiovasc. Pharmacol.*, 31, 39, 1998.

44. Broijersen, A. et al., Platelet activity in vivo in hyperlipoproteinemia — importance of combined hyperlipidemia, *Thromb. Haemost.*, 79, 268, 1998.

45. Cyrus, T. et al., Effect of low-dose aspirin on vascular inflammation, plaque stability, and atherogenesis in low-density lipoprotein receptor-deficient mice, *Circulation*, 106, 1282, 2002.

46. Pratico, D. et al., Acceleration of atherogenesis by COX-1-dependent prostanoid formation in low density lipoprotein receptor knockout mice, *Proc. Natl. Acad. Sci.*, 98, 3358, 2000.

47. Cayatte, A.J. et al., The thromboxane receptor antagonist S18886 but not aspirin inhibits atherogenesis in apo E-deficient mice: evidence that eicosanoids other than thromboxane contribute to atherosclerosis, *Arterioscler. Thromb. Vasc. Biol.*, 20, 1724, 2000.

48. Calzada, C., Bruckdorfer, K.R., and Rice-Evans, C.A., The influence of antioxidant nutrients on platelet function in healthy volunteers, *Atherosclerosis*, 128, 97, 1997.

49. Cordova, C. et al., Influence of ascorbic acid on platelet aggregation *in vitro* and *in vivo*, *Atherosclerosis*, 41, 15, 1982.

50. Bruckdorfer, K.R., Antioxidants, lipoprotein oxidation, and arterial function, *Lipids*, 31, S83, 1996.

51. de Lorgeril, M. et al., The beneficial effect of dietary antioxidant supplementation on platelet aggregation and cyclosporine treatment in heart transplant recipients, *Transplantation*, 58, 193, 1994.
52. Mabile, L., Bruckdorfer, K.R., and Rice-Evans, C., Moderate supplementation with natural alpha-tocopherol decreases platelet aggregation and low-density lipoprotein oxidation, *Atherosclerosis*, 147, 177, 1999.
53. Steiner, M., Influence of vitamin E on platelet function in humans, *J. Am. Coll. Nutr.*, 10, 466, 1991.
54. Hubbard, G.P. et al., Quercetin inhibits collagen-stimulated platelet activation through inhibition of multiple components of the glycoprotein VI signaling pathway, *J. Thromb. Haemost.*, 1, 1079, 2003.
55. Stangl, V. et al., The flavonoid phloretin suppresses stimulated expression of endothelial adhesion molecules and reduces activation of human platelets, *J. Nutr.*, 135, 172, 2005.
56. Deana, R. et al., Green tea epigallocatechin-3-gallate inhibits platelet signaling pathways triggered by both proteolytic and non-proteolytic agonists, *Thromb. Haemost.*, 89, 866, 2003.
57. Freedman, J.E. et al., Select flavonoids and whole juice from purple grapes inhibit platelet function and enhance nitric oxide release, *Circulation*, 103, 2792, 2001.
58. Petroni, A. et al., Inhibition of platelet aggregation and eicosanoid production by phenolic components of olive oil, *Thromb., Res.*, 78, 151, 1995.
59. Pace-Asciak, C.R. et al., The red wine phenolics trans-resveratrol and quercetin block human platelet aggregation and eicosanoid synthesis: implications for protection against coronary heart disease, *Clin. Chim. Acta*, 235, 207, 1995.
60. Dommimiczak, M.H., Hyperlipidemia and cardiovascular disease, *Curr. Opin. Lipidol.*, 9, 609, 1998.
61. Grundy, S.M. et al., Coordinating Committee of the National Cholesterol Education Program. Implications of recent clinical trials for the National Cholesterol Education Program Adult Treatment Panel III Guidelines, *J. Am. Coll. Cardiol.*, 44, 720, 2004.
62. Grundy, S.M., Alternative approaches to cholesterol-lowering therapy, *Am. J. Cardiol.*, 90, 1135, 2002.
63. Gotto, A.M., Jr., Management of dyslipidemia, *Am. J. Med.*, 112, 8A, 2002.
64. Gotto, A.M., Jr. and Gundy, S.M., Lowering LDL cholesterol: questions from recent meta-analyses and subset analyses of clinical trial data issues from the interdisciplinary council on reducing the risk for coronary heart disease, ninth council meeting, *Circulation*, 9, E1, 1999.
65. Gotto, A.M., Jr. and Brinton, E.A., Assessing low levels of high-density lipoprotein cholesterol as a risk factor in coronary heart disease: a working group report and update, *J. Am. Coll. Cardiol.*, 43, 717, 2004.
66. Gotto, A.M., Jr., Low high-density lipoprotein cholesterol as a risk factor in coronary heart disease: a working group report, *Circulation*, 103, 2213, 2001.
67. Castelli, W.P. et al., Lipids and risk of coronary heart disease: The Framingham Study, *Ann. Epidemiol.*, 2, 23, 1992.
68. Gordon, D.H. and Rifkind, B.M., High-density lipoprotein: the clinical implications of recent studies, *N. Engl. J. Med.*, 321, 1311, 1989.
69. Assmann, G. and Nofer, J.R., Atheroprotective effects of high-density lipoproteins, *Annu. Rev. Med.*, 54, 321, 2003.
70. Nofer, J.R. et al., HDL and arteriosclerosis: beyond reverse cholesterol transport, *Atherosclerosis*, 161, 1, 2002.

71. Brewer, H.B. et al., Regulation of plasma high-density lipoprotein levels by the ABCA1 transporter and the emerging role of high-density lipoprotein in the treatment of cardiovascular disease, *Arterioscler. Thromb. Vasc. Biol.,* 24, 1755, 2004.

72. Assmann, G. and Gotto, A.M., Jr., HDL cholesterol and protective factors in atherosclerosis, *Circulation,* 109, III8, 2004.

73. Barter, P., CETP and atherosclerosis, *Arterioscler. Thromb. Vasc. Biol., 20, 2029,* 2000.

74. Aviram, M. et al., Paraoxonase active site required for protection against LDL oxidation involves its free sulfhydryl group and is different from that required for its arylesterase/paraoxonase activities: selective action of human paraoxonase allozymes Q and R, *Arterioscler. Thromb. Vasc. Biol.,* 18, 1617, 1998.

75. Mackness, M.I. et al., Protection of low-density lipoprotein against oxidative modification by high-density lipoprotein associated paraoxonase, *Atherosclerosis,* 104, 129, 1993.

76. Mackness, M.I. and Durrington, P.N., HDL, its enzymes and its potential to influence lipid peroxidation, *Atherosclerosis,* 115, 243, 1995.

77. Pietinen, P. and Huttunen, J.K., Dietary determinants of plasma high-density lipoprotein cholesterol, *Am. Heart. J.,* 113, 620, 1987.

78. Mursu, J. et al., Dark chocolate consumption increases HDL cholesterol concentration and chocolate fatty acids may inhibit lipid peroxidation in healthy humans, *Free. Radic. Biol. Med.,* 37, 1351, 2004.

79. Nicolosi, R.J. et al., Dietary effects on cardiovascular disease risk factors: beyond saturated fatty acids and cholesterol, *J. Am. Coll. Nutr.,* 20, 421S, 2001.

80. Hannuksela, M.L., Liisanantti, M.K., and Savolainen, M.J., Effect of alcohol on lipids and lipoproteins in relation to atherosclerosis, *Crit. Rev. Clin. Lab. Sci.,* 39, 225, 2002.

81. Lavy, A. et al., Effect of dietary supplementation of red or white wine on human blood chemistry, hematology and coagulation: favorable effect of red wine on plasma high-density lipoprotein, *Ann. Nutr. Metab.,* 38, 287, 1994.

82. Navab, M. et al., The oxidation hypothesis of atherogenesis: the role of oxidized phospholipids and HDL, *J. Lipid. Res.,* 45, 993, 2004.

83. Kaplan, M. and Aviram, M., Oxidized low density lipoprotein: Atherogenic and proinflammatory characteristics during macrophage foam cell formation. An inhibitory role for nutritional antioxidants and serum paraoxonase, *Clin. Chem. Lab. Med.,* 37, 777, 1999.

84. Aviram, M., Oxidative modification of low density lipoprotein and atherosclerosis, *Isr. J. Med. Sci.,* 31, 241, 1995.

85. Aviram, M., Interaction of oxidized low density lipoprotein with macrophages in atherosclerosis and the antiatherogenicity of antioxidants, *Eur. J. Clin. Chem. Clin. Biochem.,* 34, 599, 1996.

86. Parthasarathy, S. and Rankin, S.M., The role of oxidized LDL in atherogenesis, *Prog. Lipid. Res.,* 31, 127, 1992.

87. Jialal, I. and Devaraj, S., The role of oxidized low density lipoprotein in atherogenesis, *J. Nutr.,* 126, 1053S, 1996.

88. Steinberg, D., Low density lipoprotein oxidation and its pathobiological significance, *J. Biol. Chem.,* 272, 20963, 1997.

89. Berliner, J.A. and Heinecke, J.W., The role of oxidized lipoproteins in atherosclerosis, *Free. Radic. Biol. Med.,* 20, 707, 1996.

90. Witztum, J.L. and Steinberg, D., Role of oxidized low density lipoprotein in atherogenesis, *J. Clin. Invest.,* 88, 1785, 1991.

91. Parthasarathy, S., Santanam, N., and Auge, N., Oxidized low-density lipoprotein, a two-faced janus in coronary artery disease?, *Biochem. Pharmacol.,* 56, 279, 1998.

92. Albertini, R., Moratti, R., and De Luca, G., Oxidation of low-density lipoprotein in atherosclerosis from basic biochemistry to clinical studies, *Curr. Mol. Med.,* 6, 579, 2002.

93. Herttuala, S.Y., Is oxidized low density lipoprotein present *in vivo? Curr. Opin. Lipidol.,* 9, 337, 1998.

94. Steinberg, D. et al., Beyond cholesterol: modifications of low-density lipoprotein that increase its atherogenicity, *N. Engl. J. Med.,* 320, 915, 1989.

95. Aviram, M. and Rosenblat, M., Oxidative stress in cardiovascular diseases: role of oxidized lipoproteins in macrophage foam cell formation and atherosclerosis, in *Redox Genome Interactions in Health and Disease,* Fuchs J., Podda, M., and Packer, L., eds., Marcel Dekker, New York, 2004, 557.

96. Kim, J.A. et al., Partial characterization of leukocyte binding molecules on endothelial cells induced by minimally oxidized LDL, *Arterioscler. Thromb.,* 14, 427, 1994.

97. Khan, N.B.V., Parthasarathy, S., and Alexander, R.W., Modified LDL and its constituents augment cytokine-activated vascular cell adhesion molecule-1 gene expression in human vascular endothelial cells, *J. Clin. Invest.,* 95, 1262, 1995.

98. Rangaswamy, S. et al., Exogenous oxidized low density lipoprotein injures and alters the barrier function of endothelium in rats *in vivo, Circ. Res.,* 80, 37, 1997.

99. Penn, M.S. and Chisolm, G.M., Oxidized lipoproteins, altered cell function and atherosclerosis, *Atherosclerosis,* 108, S21, 1994.

100. Witztum, J.L. and Steinberg, D., Modification of low density lipoprotein by endothelial cells involves lipid peroxidation and degradation of low density lipoprotein phospholipids, *Proc. Natl. Acad. Sci. USA,* 81, 3993, 1984.

101. Parthasarathy, S. et al., Macrophage oxidation of low density lipoprotein generates a modified form recognized by the scavenger receptor, *Arteriosclerosis,* 6, 505, 1986.

102. Kaplan, M. and Aviram, M., Macrophage-mediated oxidation of LDL and atherogenesis: protective role for paraoxonases, in *Cellular Dysfunction in Atherosclerosis and Diabetes — Reports from Bench to Bedside,* Simionesco, M., Sima, A., and Popov, D., eds., Romanian Academy Publishing, Romania, 2004, 336.

103. Aviram, M. and Rosenblat, M., Macrophage mediated oxidation of extracellular low density lipoprotein requires an initial binding of the lipoprotein to its receptor, *J. Lipid. Res.,* 35, 385, 1994.

104. Aviram, M. et al., Activation of NADPH oxidase is required for macrophage-mediated oxidation of low density lipoprotein, *Metabolism,* 45, 1069, 1996.

105. Aviram, M. and Fuhrman, B., LDL oxidation by arterial wall macrophages depends on the antioxidative status in the lipoprotein and in the cells: role of prooxidants vs. antioxidants, *Mol. Cell. Biochem.,* 188, 149, 1998.

106. Fuhrman, B., Oiknine, J., and Aviram, M., Iron induces lipid peroxidation in cultured macrophages, increases their ability to oxidatively modify LDL and affect their secretory properties, *Atherosclerosis,* 111, 65, 1994.

107. Fuhrman, B. et al., Increased uptake of low density lipoprotein (LDL) by oxidized macrophages is the result of enhanced LDL receptor activity and of progressive LDL oxidation, *Free. Radic. Biol. Med.,* 23, 34, 1997.

108. Jacob, R.A., Evidence that diet modification reduces *in vivo* oxidant damage, *Nutr. Rev.,* 57, 255, 1999.

109. Aviram, M., Macrophage foam cell formation during early atherogenesis is determined by the balance between pro-oxidants and anti-oxidants in arterial cells and blood lipoproteins, *Antioxid. Redox. Signal.,* 1, 585, 1999.

110. Halliwell, B., Free radicals, antioxidants and human disease: Curiosity, cause, or consequence, *Lancet,* 344, 721, 1994.

111. Stocker, R., Dietary and pharmacological antioxidants in atherosclerosis, *Curr. Opin. Lipidol.,* 10, 589, 1999.

112. Futterman, L.G. and Lemberg, L., The use of antioxidants in retarding atherosclerosis: fact or fiction? *Am. J. Crit. Care.,* 8, 130, 1999.

113. Gaziano, J.M., Antioxidant vitamins and cardiovascular disease, *Proc. Assoc. Am. Physicians.,* 111, 2, 1999.

114. Fuhrman, B. and Aviram, M., Antiatherogenicity of nutritional antioxidants, *Idrugs,* 4, 82, 2001.

115. Aviram, M. et al., Dietary antioxidants against LDL oxidation and atherosclerosis development: protective role for paraoxonase, in *Handbook of Experimental Pharmacology (HEP): Arteriosclerosis; Influence of Diet and Drugs,* von Eckardstein, A., ed., 2005, 259.

116. Maor, I. et al., Plasma LDL oxidation leads to its aggregation in the atherosclerotic apolipoprotein E-deficient mice, *Arterioscler. Thromb. Vasc. Biol.,* 17, 2995, 1997.

117. Cyrus, T. et al., Vitamin E reduces progression of atherosclerosis in low-density lipoprotein receptor-deficient mice with established vascular lesions, *Circulation,* 107, 521, 2003.

118. Rice-Evans, C.A. et al., The relative antioxidant activities of plant-derived polyphenolic flavonoids, *Free. Radic. Res.,* 22, 375, 1995.

119. Van Acker, S.A.B.E. et al., Structural aspects of antioxidants activity of flavonoids, *Free Radic. Biol. Med.,* 20, 331, 1996.

120. Fuhrman, B. and Aviram, M., Flavonoids protect LDL from oxidation and attenuate atherosclerosis, *Curr. Opin. Lipidol.,* 12, 41, 2001.

121. Fuhrman, B. and Aviram, M., Polyphenols and flavonoids protect LDL against atherogenic modifications, in *Handbook of Antioxidants Biochemical, Nutritional and Clinical Aspects, 2nd Edition.* Cadenas, E., and Packer, L., eds., Marcel Dekker, New York, 2001, 303.

122. Rosenblat, M. et al., Macrophage enrichment with the isoflavan glabridin inhibits NADPH oxidase-induced cell-mediated oxidation of low density lipoprotein: A possible role for protein kinase C, *J. Biol. Chem.,* 274, 13790, 1999.

123. Hertog, M.G. et al., Flavonoid intake and long-term risk of coronary heart disease and cancer in the seven countries study, *Arch. Intern. Med.,* 155, 381, 1995.

124. Knekt, P. et al., Flavonoid intake and risk of chronic diseases, *Am. J. Clin. Nutr.,* 76, 560, 2002.

125. Knekt, P. et al., Flavonoid intake and coronary mortality in Finland: a cohort study, *Brit. Med. J.,* 312, 478, 1996.

126. Renaud, S. and de Lorgeril, M., Wine alcohol, platelets and the French paradox for coronary heart disease, *Lancet,* 339, 1523, 1992.

127. Aviram, M., Antioxidants in restenosis and atherosclerosis, *Curr. Interven. Cardiol. Rep.,* 1, 66, 1999.

128. Hayek, T. et al., Reduced progression of atherosclerosis in the apolipoprotein E-deficient mice following consumption of red wine, or its polyphenols quercetin, or catechin, is associated with reduced susceptibility of LDL to oxidation and aggregation, *Arterioscler. Thromb. Vasc. Biol.,* 17, 2744, 1997.

129. Aviram, M. and Fuhrman, B., Wine flavonoids, LDL cholesterol oxidation and atherosclerosis, in *Wine: a scientific exploration,* Sandler, M. and Pinder, R.M., eds., Taylor & Francis, London, 2003, 140.

130. Fuhrman, B. et al., Licorice extract and its major polyphenol glabridin protect low density lipoprotein against lipid peroxidation: *in vitro* and *ex vivo* studies in humans and in atherosclerotic apolipoprotein E-deficient mice, *Am. J. Clin. Nutr.,* 66, 267, 1997.

131. Aviram, M., Vaya, J., and Fuhrman, B., Licorice root flavonoid antioxidants reduce LDL oxidation and attenuate cardiovascular diseases, in *Herbal Medicines: Molecular Basis of Biological Activity and Health*, Packer, L., Halliwel, B., and Nam Ong C., eds., Marcel Dekker, New York, 2004, 595.

132. Fuhrman, B. et al., Grape powder polyphenols attenuate atherosclerosis development in apolipoprotein E deficient (E^0) mice and reduce macrophage atherogenicity, *J. Nutr.*, 135, 722, 2005.

133. Fuhrman, B. et al., Ginger extract consumption reduces plasma cholesterol levels, inhibits LDL oxidation and attenuates development of atherosclerosis in the atherosclerotic apolipoprotein E deficient mice, *J. Nutr.*, 130, 1124, 2000.

134. Langley, P., Why a pomegranate? *Brit. Med. J.*, 321, 1153, 2000.

135. Aviram, M., Polyphenols from pomegranate juice, red wine and licorice root protect against lipids peroxidation and attenuate cardiovascular diseases, in *Polyphenols 2000. XXth International Conference on Polyphenols*, Martens, S., Treutter, D. and Forkmann, G., eds., Freising-Weihenstephan, Germany, 2002, 158.

136. Gil, M.I. et al., Antioxidant activity of pomegranate juice and its relationship with phenolic composition and processing, *J. Agric. Food Chem.*, 48, 4581, 2000.

137. Ben Nasr, C., Ayed, N., and Metche, M., Quantitative determination of the polyphenolic content of pomegranate peel, *Z. Lebensm. Unters. Forsch.*, 203, 374, 1996.

138. Aviram, M., Pomegranate juice as a major source for polyphenolic flavonoids and it is most potent antioxidant against LDL oxidation and atherosclerosis, in *Proceedings of the 11th Biennal Meeting of the Society for Free Radical Research International*, Monduzzi, S.p.A., ed., MEDIMOND, Paris, 2002, 523.

139. Aviram, M. et al., Pomegranate juice polyphenols decreases oxidative stress, low-density lipoprotein atherogenic modifications and atherosclerosis, *Free Radic. Res.*, 36, (Supplement 1), 72, 2002.

140. Aviram, M. et al., Pomegranate juice flavonoids inhibit low-density lipoprotein oxidation and cardiovascular diseases: studies in atherosclerotic mice and in humans, *Drugs Ex. Clin. Res.*, 28, 49, 2002.

141. Cerda, B. et al., Evaluation of the bioavailability and metabolism in the rat of punicalagin, an antioxidant polyphenol from pomegranate juice, *Eur. J. Nutr.*, 42, 18, 2003.

142. Cerda, B. et al., Repeated oral administration of high doses of the pomegranate ellagitannin punicalagin to rats for 37 days is not toxic, *J. Agric. Food Chem.*, 51, 3493, 2003.

143. Seeram, N.P., Lee, R., and Heber, D., Bioavailability of ellagic acid in human plasma after consumption of ellagitannins from pomegranate (Punica granatum L.) juice, *Clin. Chim. Acta*, 348, 63, 2004.

144. Perez-Vicente, A., Gil-Izquierdo, A., and Garcia-Viguera, C., *In vitro* gastrointestinal digestion study of pomegranate juice phenolic compounds, anthocyanins, and vitamin C, *J. Agric. Food Chem.*, 50, 2308, 2002.

145. Aviram, M. et al., Pomegranate juice consumption reduces oxidative stress, atherogenic modifications to LDL, and platelet aggregation: studies in humans and in the atherosclerotic apolipoprotein E-deficient mice, *Am. J. Clin. Nutr.*, 71, 1062, 2000.

146. Kaplan, M. et al., Pomegranate juice supplementation to atherosclerotic mice reduces macrophages lipid peroxidation, cellular cholesterol accumulation and development of atherosclerosis, *J. Nutr.*, 131, 2082, 2001.

147. de Nigris, F. et al., Beneficial effects of pomegranate juice on oxidation-sensitive genes and endothelial nitric oxide synthase activity at sites of perturbed shear stress, *Proc. Natl. Acad. Sci. U.S.A.*, 102, 4896, 2005.

148. Aviram, M. et al., Pomegranate juice consumption for 3 years by patients with carotid artery stenosis reduces common carotid intima-media thickness, blood pressure and LDL oxidation, *Clin. Nutr.*, 23, 423, 2004.

149. Kitiyakara, C. and Wilcox, C.S., Antioxidants for hypertension, *Curr. Opin. Nephrol. Hypertens.*, 7, 531, 1998.

150. Duffy, S.J. et al., Treatment of hypertension with ascorbic acid, *Lancet*, 354, 2048, 1999.

151. Galley, H.F., Thornton, J., Howdle, P.D., Walker, B.E., and Webster, N.R., Combination oral antioxidant supplementation reduces blood pressure, *Clin. Sci.*, 92, 361, 1997.

152. Singh, R.B. et al., Effect of hydrosoluble coenzyme Q10 on blood pressures and insulin resistance in hypertensive patients with coronary artery disease, *J. Hum. Hypertens.*, 13, 203, 1999.

153. Aviram, M. and Dornfeld, L., Pomegranate juice consumption inhibits serum angiotensin converting enzyme activity and reduces systolic blood pressure, *Atherosclerosis*, 158, 195, 2001.

154. He, K. et al., Inactivation of cytochrome P-450 3A4 by bergamottin, a component of grapefruit juice, *Chem. Res. Toxicol.*, 11, 252, 1998.

155. Aviram, M., Kent, U.M., and Hollenberg, P.F., Microsomal cytochromes P450 catalyze the oxidation of low density lipoprotein, *Atherosclerosis*, 143, 253, 1999.

156. Polagruto, J.A. et al., Effects of flavonoid-rich beverages on prostacyclin synthesis in humans and human aortic endothelial cells: association with ex vivo platelet function, *J. Med. Food*, 6, 301, 2003.

157. Esmaillzadeh, A. et al., Concentrated pomegranate juice improves lipid profiles in diabetic patients with hyperlipidemia, *J. Med. Food*, 7, 305, 2004.

158. Chidambara Murthy, K.N., Jayaprakasha, G.K., and Singh, R.P., Studies on antioxidant activity of pomegranate (*Punica granatum*) peel extract using *in vivo* models, *J. Agric. Food Chem.*, 50, 4791, 2002.

159. Aviram, M. et al., Human serum paraoxonase (PON1) is inactivated by oxidized low density lipoprotein and preserved by antioxidants, *Free Radic. Biol. Med.*, 26, 892, 1999.

160. Berliner, J.A. et al., Atherosclerosis: basic mechanisms: Oxidation, inflammation and genetics, *Circulation*, 9, 2488, 1995.

161. Fuhrman, B., Volkova, N., and Aviram, M., Oxidative stress increases the expression of the CD36 scavenger receptor and the cellular uptake of oxidized low-density lipoprotein in macrophages from atherosclerotic mice: protective role of antioxidants and of paraoxonase, *Atherosclerosis*, 161, 307, 2002.

162. Aggarwal, B.B. and Shishodia, S., Suppression of the nuclear factor κB activation pathway by spice-derived phytochemicals: reasoning for seasoning, *Ann. N.Y. Acad. Sci.*, 1030, 434, 2004.

163. Schubert, S.Y., Neeman, I., and Resnick, N., A novel mechanism for the inhibition of NF-κB activation in vascular endothelial cells by natural antioxidants, *FASEB J.*, 16, 1931, 2002.

164. Schubert, S.Y., Lansky, E.P., and Neeman, I., Antioxidant and eicosanoid enzyme inhibition properties of pomegranate seed oil and fermented juice flavonoids, *J. Ethnopharmacol.*, 66, 11, 1999.

165. Goldstein, J.L. and Brown, M.S., Regulation of the mevalonate pathway, *Nature*, 343, 425, 1990.

166. Brown, M.S. and Goldstein, J.L., A receptor-mediated pathway for cholesterol homeostasis. *Science*, 232, 34, 1986.

167. Krieger, M., The best of cholesterols, the worst of cholesterols: a tale of two receptors, *Proc. Natl. Acad. Sci. USA*, 4, 4077, 1998.

168. Fuhrman, B., Volkova, N., and Aviram, M., Pomegranate juice inhibits oxidized LDL uptake and cholesterol biosynthesis in macrophages, *J. Nutr. Biochem.*, 16, 570, 2005.

169. Rosenblat, M., Hayek, T., and Aviram, M. Anti-oxidative effects of pomegranate juice (PJ) consumption by diabetic patients on serum and on macrophages. *Atherosclerosis* 2006 (in press).

170. Rosenblat, M., Volkova, N., Coleman, R., and Aviram, M. Pomegranate byproduct administration to apolipoprotein e-deficient mice attenuates atherosclerosis development as a result of decreased macrophage oxidative stress and reduced cellular uptake of oxidized low-density lipoprotein. *J. Agric. Food Chem.* 54, 1928–1935, 2006.

5 Protection against Stroke

Marva I. Sweeney-Nixon

CONTENTS

5.1 INTRODUCTION

Cardiovascular diseases (CVD) such as heart attack and stroke are the number one cause of mortality in North America[1-3] and indeed many parts of the world.[4-6] According to the World Health Organization (WHO), "By 2010, CVD will be the leading cause of death in developing countries."[6] Not everyone who suffers a heart attack or stroke will die; the WHO estimates that worldwide, at least 20 million people survive heart attacks and strokes every year, thus requiring medical care that is very costly.[6] Stroke is especially debilitating since it is the leading cause of serious long-term disability.[1]

Direct costs (e.g., expenses related to hospital stays and drugs) and indirect costs (e.g., long-term disabilities and premature death) incurred as a result of cardiovascular disease are in excess of $360 billion annually in the U.S.[1] In Canada, total annual costs attributable to cardiovascular diseases (approximately $20 billion) are higher than those incurred by cancer, injuries, and all other categories.[3] Unfortunately, cardiovascular diseases are rarely cured and are merely managed by medical and surgical treatments.[3] More often than not, the underlying disease progresses and worsens, such that the burden on the patient, the family, and the health care system accumulates.[6] Because of this, the WHO, the Heart and Stroke Foundation of Canada, and The Centers for Disease Control and Prevention (among several other organizations, associations, and coalitions) are promoting a heart disease and stroke prevention strategy.[3,6]

Much research in the last 15 years has shown that consumption of fruits and vegetables plays a role in preventing the onset and slowing the progression of heart disease and stroke.[7-10] Young vegetarians, who did not eat any meat or fish, have been shown to have a 24% lower risk of dying of ischemic heart disease,[11] which is not surprising since plant-based foods are low in saturated fat and cholesterol. However, strong evidence points to the content of polyphenols in plant-based foods and beverages such as tea and wine as being correlated with cardiovascular protection.[12-14]

This chapter will review the literature that demonstrates the efficacy of dietary consumption of pomegranates, rich in flavonoids and other antioxidant phytonutrients,[15-17] in promoting good cardiovascular health, with an emphasis on stroke.

5.2 CARDIOVASCULAR DISEASES

5.2.1 TYPES OF CARDIOVASCULAR DISEASES

Any disease or injury to the cardiovascular system (which encompasses the heart and blood vessels) is considered a cardiovascular disease. Disorders of the heart account for most cardiovascular diseases, and include ischemic heart disease, heart attack, and congestive heart failure. Diseases of the vascular system, namely to arteries and veins that supply and remove blood from all tissues, include hypertension and stroke.[1-3]

5.2.1.1 Heart Disease

Ischemic heart disease (coronary artery disease) is the most common spectrum of heart disease.[1-6] It develops gradually as the coronary arteries supplying blood to the heart become progressively blocked with atherosclerotic plaque, a combination of oxidized low-density lipoproteins (LDL), inflammatory cells (macrophages), smooth muscle cells, and debris[12,18] (see other chapters). Eventually, blood supply may become completely interrupted by plaque, or more commonly by a coronary thrombosis (blood clot) when plaque ruptures, precipitating a heart attack (myocardial infarction, MI).[18] As ischemia continues for several minutes, the myocardium is deprived of oxygen and muscle cells may die, leading to permanent damage to heart tissue, heart failure, or sudden death. Other heart disorders include arrhythmias (dysfunction of the electrical conduction system of the heart); valve disorders; and congestive heart failure. The pain or discomfort associated with advanced ischemic heart disease is known as angina.

5.2.1.2 Stroke

Stroke is a cerebrovascular accident, characterized by a reduction in blood supply to a region of the brain. This leads to a reduction of oxygen tension and of high-energy metabolites in an area of the brain. Previously held to be a disease of the elderly, cerebrovascular compromise is now recognized as affecting all age groups. There are two main types of stroke, namely ischemic (thrombotic) stroke and hemorrhagic stroke.[19] As the name implies, a thrombotic stroke occurs when there is a clot, embolus or other blockage, such as narrowing induced by atherosclerotic plaque, in a cerebral artery.[19] This causes a sudden cessation of blood flow (ischemia) that is usually temporary, followed by dislodging of the clot and reestablishment of blood flow (reperfusion). As brain tissue is deprived of oxygen and nutrients, neuronal cells start to die, leading to long-term deficits in brain function (see Section 5.5). Approximately 80% of strokes are ischemic.[19]

Rupture of a blood vessel that is weakened (e.g., when an aneurysm occurs), brittle (after years of hypertension and atherosclerosis), or defective (when arteriovenous malformation is present) leads to hemorrhagic stroke.[19] Within this category, differences exist. For example, a brain hemorrhage can be intracerebral (within the brain; intracranial hematoma) or subarachnoid (between the brain and the overlying skull and connective tissue layers), depending on the site of the ruptured blood vessel. Regardless of the site, the leaking of blood into and around the brain increases intracranial pressure, which compresses brain tissue. As well, the flow of blood to certain brain regions is disrupted, ultimately damaging the brain.[19]

5.3 CARDIOVASCULAR DISEASE STATISTICS

The 2002 World Report on Violence and Health published by the WHO (Geneva)[4] lists ischemic heart disease as the top cause of death among its member states, accounting for 12.4% of all deaths in 2000.[4,5] Cerebrovascular disease, or stroke, accounted for 9.2% of deaths.[4] In Canada, cardiovascular disease accounted for 74,626 Canadian deaths (32% of all male deaths and 34% of all female deaths) in 2002.[2,3]

In the U.S.[1,19] and Canada,[2,3] stroke is the third leading cause of death after heart disease and cancer, accounting for close to 7% of all deaths. Each year approximately 500,000 Americans suffer a first stroke and another 200,000 experience a recurrent attack.[1] Demographically, more women than men have a stroke; increasing age and race (African Americans) increases the chance of suffering a stroke. Of the two types of stroke, 8 to 12% of ischemic strokes and 37 to 38% of hemorrhagic strokes result in death within 30 days, while one quarter of all stroke victims die within a year. Of the survivors, 14% will have another stroke (or TIA) within a year.[1]

5.4   RISK FACTORS FOR CARDIOVASCULAR DISEASES

There are several factors that increase the risk of developing CVD. These risk factors may be nonmodifiable, such as a family history of cardiovascular disease, race (black, aborginal), and age.[1,3,19] However, many of the risk factors for heart attack and stroke are indeed modifiable with changes in behavior, drugs, or diet. Eighty percent of the Canadian population has at least one modifiable risk factor for cardiovascular disease.[3] Smoking is "the single most important cause of preventable illness and premature death for Canadians."[3] Other risk factors that can be controlled are diabetes and obesity (or metabolic syndrome[20,21]), sedentary lifestyles, stress, hypertension,[3,10] and dyslipidemia.[3,8] Thus, the incidence of cardiovascular disease can be attenuated by increases in physical activity, smoking cessation, controlling diabetes with insulin or oral hypoglycemic agents, controlling hypertension with various antihypertensive drugs, cholesterol-lowering therapy with drugs, and better choices in diet. For example, the Dietary Approaches to Stop Hypertension (DASH) diet, which includes lots of fruits, vegetables, nuts, legumes, and low-fat dairy products, while limiting meat, sugars, fats, and salt, is a viable way to prevent and treat hypertension.

5.5   MECHANISMS OF STROKE DAMAGE

Ischemic stroke is thought to be the product of atherosclerotic plaque inflammation followed by plaque rupture and thrombosis (with or without embolism), which occludes cerebral arteries.[19,22–25] Platelet activation is a crucial mechanism in arterial thrombogenesis and therefore in the pathophysiology of ischemic stroke.[23] Accordingly, antiplatelet therapy plays a central role in prevention of ischemic stroke. Despite many experimental studies, there is as yet no generally accepted effective treatment in the acute phase of stroke.

Neuronal injury after an occlusion of a cerebral artery can be loosely divided into three phases: acute, which occurs within hours of occlusion; subacute, occurring hours to days after the insult; and chronic, which lasts anywhere from several days to even months.[24]

5.5.1   ACUTE PHASE OF NEURONAL INJURY

Acutely, neuronal cells deprived of oxygen during arterial thrombosis cannot produce sufficient ATP. This leads to a complete loss of homeostasis that induces sudden necrotic cell death via several different mechanisms (reviewed extensively elsewhere [e.g., References 25, 26]). Calcium ion influx into hypoxic cells is a final common pathway, leading to cell death by activation of Ca^{2+}-dependent proteases, increases in osmotic pressure, and subsequent cellular swelling, and dysregulation of the mitochondria, resulting in even less ATP production and an increase in production of reactive oxygen species (ROS).[26] Oxidative stress may then up-regulate extracellular signal regulated kinases and, subsequently, the arachidonic acid cascade mediated by phospholipase A2 and cyclooxygenase-2 (COX-2) leading to neurodegenerative changes in the brain.[26] These events occur at the core of the infarction. The area of damage around the core is referred to as the ischemic penumbra[25] — it suffers milder insults that involve both mild excitotoxic and inflammatory mechanisms, leading to delayed cell death, which shows characteristics of apoptosis.[27]

Necrotic neurons release tumor necrosis factor-α and interleukin-1β, which induce further expression of transmembrane glycoproteins (e.g., E-selectin) on vascular endothelial cells in the brain. E-selectin attracts a massive influx of leukocytes, especially neutrophils, into the ischemic region. Neutrophils adhere to the endothelium and release superoxide and other ROS, the predominant mediators of reperfusion injury.[28-31] ROS damage neurons by inducing lipid peroxidation, or oxidizing protein and nucleic acids.[26,32] ROS are also involved in cell death/survival signaling pathways in the chronic phase.[33] Leukocyte infiltration also releases neurotoxins in the ischemic brain, which exacerbates the acute necrotic damage and promotes microvascular occlusion in the ischemic penumbra.[29,31]

5.5.2 SUBACUTE PHASE OF BRAIN INJURY

During the subacute phase (hours after ischemia), the migration of leukocytes into the compromised brain tissue (Section 5.5.1) amplifies inflammatory signaling cascades.[24,34-36] Inflammatory mediators and proteases such as the matrix metalloproteinases (MMP-2 and -9) and interleukins (IL-1) are thought to mediate damage during the inflammatory phase of ischemia.[24,37,38] Indeed, inhibitors of MMP-9 are neuroprotective in experimental stroke, and MMP-9 knockout mice show much smaller ischemic lesions.[38] Proinflammatory genes such as COX-2 and IL-1β are upregulated in the ischemic brain;[35,36] COX-2 activity has been implicated in the pathogenesis of neuronal cell death in ischemia and other diseases, but the mechanism by which COX-2 exacerbates cell death is unknown.

5.5.3 CHRONIC PHASE OF BRAIN INJURY

Cumulative evidence suggests that apoptosis contributes to ischemic injury in the brain, especially in the neurons of the penumbra.[27,32,33,39,40] During the reperfusion that follows transient focal and global ischemia, reoxygenation provides a substrate for numerous oxidation reactions. The ROS produced by the mitochondria cause initial necrotic death to cells[26] but recent evidence suggests that ROS also initiate cell signaling pathways after cerebral ischemia,[40] which contributes to the pathogenesis of apoptotic cell death in ischemic lesions. Proapoptotic gene products such as the caspases, Bax, and Trp53 predominate, and the antiapoptotic proteins Bcl2 and IAP are protective.[24,40] In addition, VEGF may promote angiogenesis, and superoxide dismutases may protect against further oxidative damage.[40] Eventually, the injured blood vessel may either promote recovery by angiogenesis, underapoptosis, or atherosclerosis, depending on its size and function.[24]

5.6 POMEGRANATES

5.6.1 CHEMISTRY

Pomegranates *(Punica granatum)* are an excellent dietary source of polyphenolic antioxidants. In a comparative study, pomegranates were shown to contain 11.33 mmol total antioxidants/100g vs. 8.23 mmol antioxidants/100 g in wild bilberry *(Vaccinium myrtillus)*, 2.42 mmol antioxidants/100 g for grapes *(Vitis vinifera)* and 0.34 mmol antioxidants/100 g for tomatoes *(Lycopersicon esculentum)*.[17] In North America, most pomegranate consumption occurs from drinking pomegranate juice. The total polyphenol concentration in pomegranate juice is higher than that of green tea and red wine, giving pomegranate juice a threefold higher antioxidant capacity.[15] Commercial pomegranate juice contains over 20 different polyphenols: anthocyanins (delphinidin, cyanidin, and pelargonidin and their glycosides), hydrolyzable tannins (ellagitannins, gallotannins, punicalagin, punicalin), and phenolics (ellagic acid, gallic acid);[15,16] other than those named, most have not yet been identified. Other components of note are water soluble poly- and oligo-saccharides (containing mainly glucose, sucrose, xylose, and galactose compounds), organic acids (citric, malic), and seed oil lipids (punicic acid). Ascorbic acid (vitamin C) is found in pomegranate fruits but is destroyed during pasteurization and therefore is not present in pasteurized pomegranate juice.

5.6.2 EFFECTS OF POMEGRANATES ON ANTIOXIDANT STATUS

Pomegranate anthocyanins have been shown to scavenge superoxide ($O_2^{\cdot-}$) and hydroxyl ($OH^{\cdot-}$) radicals to prevent lipid peroxidation in rat brain homogenates.[16] An extract from pomegranate peel fed to rats for 14 days reduced lipid peroxidation in the liver as well.[41] The plasma antioxidant status of humans fed pomegranate juice is elevated over control subjects,[42–44] suggesting that pomegranate polyphenols are bioavailable and are able to elevate the antioxidant capacity of the body. Pomegranate polyphenols have been shown to increase the concentrations of the endogenous antioxidant glutathione (GSH) in mouse peritoneal macrophages,[42] while pomegranate peel extract, fed to rats for 14 days, prevented the carbon tetrachloride-induced loss of peroxidase, catalase, and superoxide dismutase in the liver.[41]

Recently, we fed rats three different amounts of pomegranate juice (0.2 to 40 μmol total polyphenols/day) and measured brain concentrations of reduced and oxidized glutathione (GSH and GSSG respectively), using reverse-phase HPLC with UV detection.[45,46] Higher concentrations of GSH were measured in the brain of rats fed pomegranate juice when compared to rats drinking water (Figure 5.1). The effect was time dependent, with the greatest effects (2.9-fold increase over control) seen after 6 weeks of diet (Figure 5.1C). There were no apparent diet-related differences in the amount of oxidized glutathione (GSSG) in the brain (Figure 5.1D to F). Thus, the ratios of reduced-to-oxidized forms of glutathione were elevated by pomegranate diets in a dose- and time-dependent manner (Figure 5.2). The redox state of cells is most accurately determined by measuring the ratios of the reduced-to-oxidized forms of various redox couples, such as GSH:GSSG.[47] Thus, pomegranate juice consumption improves the redox state of the brain.

We also found an increase in the activity of GSH metabolizing enzymes. The activities of glutathione peroxidase (EC 1.11.1.9), glutathione reductase (EC 1.8.1.7), and glutathione-S-transferase (EC 2.5.1.18) were determined in brain homogenates using commercially available spectrophotometric assay kits (Cayman Chemical Company, Ann Arbor, MI). Feeding pomegranate juice to rats for 2, 4, or 6 weeks

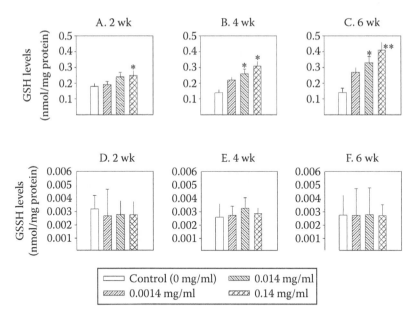

FIGURE 5.1 Effect of drinking pomegranate polyphenols for 2, 4, or 6 weeks, on the concentrations of reduced glutathione (GSH, panels A–C) and oxidized glutathione (GSSG, panels D-F) in rat brains. Data are means ± SEM of nmol GSH or GSSG per mg of protein (N = 7–8 per group). *, $p < 0.05$, ** $p < 0.01$, ANOVA with Tukey's compared to 0 mg/ml.

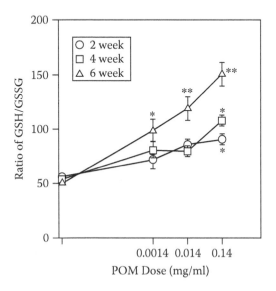

FIGURE 5.2 Effect of drinking pomegranate polyphenols (POM) for 2, 4, or 6 weeks, on the ratios of reduced to oxidized glutathione (GSH:GSSH) in rat brains.

TABLE 5.1
Effect of Drinking Pomegranate Polyphenols on Brain Glutathione Peroxidase Activity

Concentration of Pomegranate Polyphenols Fed (mg/ml)	Brain Glutathione Peroxidase Activity (nmol NADPH oxidized/mg protein/min)		
	2-Week Feeding	4-Week Feeding	6-Week Feeding
0	102.0 ± 21.5[a]	94.0 ± 14.0	109.5 ± 15.5
0.0014	116.5 ± 7.2	133.1 ± 33.0	161.2 ± 16.0[b]
0.014	123.0 ± 26.1	160.5 ± 19.0[b]	252.5 ± 33.6[b]
0.14	147.5 ± 8.6[b,c]	202.0 ± 26.0[b]	309.0 ± 28.5[b,c]

[a] Mean ± SEM for all such values, N = 7–8 rats per treatment group
[b] $p < 0.05$, ANOVA with Tukey's posthoc test compared to respective control (0 mg/ml)
[c] $p < 0.05$, ANOVA with Tukey's posthoc test compared to 4 week

increased glutathione peroxidase activity (Table 5.1). The effects at 2 weeks were only evident in rats receiving the highest dose of polyphenols, while the effects at 4 and 6 weeks were evident at lower doses (Table 5.1). However, the activity of brain glutathione-S-transferase and glutathione reductase was not different among any diet groups except at the highest dose of polyphenols in the 4-wk treatment group (data not shown).

GSH is the major antioxidant system in cells, donating electrons to H_2O_2 and becoming oxidized to GSSG in a reaction catalyzed by glutathione peroxidase. Intracellular antioxidant content, especially glutathione concentrations, are thought to be a key factor in determining neuronal vulnerability to oxidative stress.[48] Treatments that increase GSH have been shown to reduce oxidative stress[46,49–51] and glutathione peroxidase over-expression protects the brain from ischemic damage.[49] Thus, our results show for the first time that pomegranate consumption enhances the content of brain GSH and the activity of brain glutathione peroxidase, which may play a key role in protecting the brain from oxidative stress. Pomegranate diets may have induced an increase in GSH synthesis via increase in the synthesis of γ-glutamylcysteine synthetase, as has been seen for quercetin, kaempferol, apigenin, and ellagic acid.[52,53] During oxidative stress, GSH is oxidized to GSSG, which may react with cysteinyl thiols of cellular proteins to form protein-glutathione disulfides, a reaction known as S-thiolation or glutathiolation, among the most sensitive and physiologically relevant cellular metabolic events occurring during oxidative stress.[54] In this case an increase in GSSG would not be expected.

5.6.3 EFFECTS OF POMEGRANATES ON ATHEROSCLEROSIS AND HYPERTENSION

Atherosclerosis is a major risk factor for thrombotic stroke and hypertension is a risk factor for both types of strokes. Pomegranate juice has a profound effect on reducing the onset and progression of atherosclerosis, as discussed in pertinent chapters in this book. Briefly, atherosclerotic mice fed pomegranate juice had a slower accumulation of blood cholesterol and the development of atherosclerosis was delayed.[42,55,56] Similar results in humans were recently published showing a 30% decrease in carotid artery intima-media thickness after consumption of pomegranate juice for 1 year.[57] Pomegranate juice consumption was also shown to reduce systolic blood pressure in hypertensive[58] and atherosclerotic[57] patients, possibly due to decreased serum angiotensin converting enzyme activity.[58] When the effects of pomegranate on platelet activity was evaluated in humans, prolonged clotting times were observed after feeding healthy adults small amounts of pomegranate juice daily for 5 days.[59]

5.6.4 Effects of Pomegranates on Stroke

5.6.4.1 Neonatal Animals

Until recently, little was known about the effects of pomegranate consumption or isolated pomegranate polyphenols on hypoxia-ischemia (H-I) damage in the brain. In March 2005, Loren et al.[60] published a study demonstrating the neuroprotective effect of pomegranate juice, fed to mouse dams during pregnancy and lactation, on hypoxic-ischemic brain damage in neonatal mice. Dams were fed pomegranate juice (8 to 32 μmol/day of total polyphenols) for 15 to 21 days. Pups on postnatal day 7 were used to induce unilateral H-I by ligation of the left common carotid artery, followed 2 hr later by exposure to 8% oxygen. One day after H-I, caspase-3 activity was elevated in the hippocampal and cortical areas of the brain on the hemisphere ipsilateral to the ligation. Pups whose dams had been fed pomegranate juice had 64 to 84% less induction of caspase-3, a marker of apoptosis.[60] This translated into smaller lesions in the hippocampus, cortex, and striatum 7 days after H-I. The group that was fed the highest dose of pomegranate juice had greater than 50% neuroprotection.[60] Thus, maternal supplementation with pomegranate juice is neuroprotective to the neonatal brain when challenged with hypoxia and ischemia, at least in mice.[60]

5.6.4.2 Adult Animals

Using an adult rat model of brain H-I,[61] we have recently observed a similar beneficial effect of diets containing pomegranate juice on ischemic stroke (Sweeney, M.I., Harmon, M.I., Durant, C.D., Soloman, F., and Schulman, R.N., unpublished). Pomegranate juice concentrate from the Wonderful variety of pomegranates (Pom Wonderful LLC, Los Angeles, CA) containing 8.82 mg total polyphenols/ml was serially diluted in distilled water. The resulting three solutions (0.0014, 0.014, or 0.14 mg/ml of total pomegranate polyphenols in water) were offered to male rats to drink *ad libitum* for 2, 4, or 6 weeks, while being fed normal rat chow. A molar dose of total polyphenols was estimated at 0.2, 2.4, and 30 μmol/day using an average molecular weight for dietary polyphenols (350 g/mol)[42,55] and the actual amount of fluid consumed daily. Brain damage was induced surgically by unilateral carotid artery occlusion followed by 8% oxygen (hypoxia) 24 hours later, and assessed histologically. A dramatic reduction in H-I damage (by 33 to 93%) to the hippocampus was seen (Figure 5.3) when rats drank pomegranate polyphenols, dependent

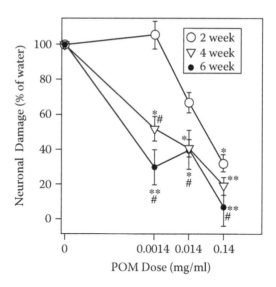

FIGURE 5.3 Effect of drinking pomegranate polyphenols (POM) for 2, 4, or 6 weeks, on the severity of neuronal damage in the hippocampus induced by unilateral hypoxia-ischemia. *$p < 0.05$, **$p < 0.01$, 2-way ANOVA with Tukey's post hoc analysis compared to control. #$p < 0.05$, 2-way ANOVA with Tukey's post hoc analysis compared to 2 wk (n = 7–8).

on dose ingested and duration of feeding. The longer the feeding period and the higher the dose of ingested polyphenols, the more profound the neuroprotection.

Rats whose diet had been supplemented for 2 weeks with the highest dose of pomegranate polyphenols (0.14 mg/ml) had 68% fewer cells damaged than in the brains of control animals (Figure 5.3). There was no statistically significant neuroprotection when lower doses of pomegranate polyphenols were ingested for 2 weeks. All groups of animals drinking pomegranate juice for 4 weeks experienced neuroprotection (48 to 81% less ischemic damage) that was dose-dependent. The effect seen at the lowest dose at 4 weeks was greater than that seen at 2 weeks ($p < 0.05$, two-way ANOVA with Tukey's posthoc analysis), indicating a time-dependent effect. Similarly, all treatment groups fed pomegranate juice for 6 weeks experienced neuroprotection in ischemic damage that was dose-dependent and time-dependent. The effects at the lowest dose were greater than those at 2 weeks and 4 weeks, and all the effects at 2 weeks were statistically different than the effect at 6 weeks ($p < 0.05$, two-way ANOVA with Tukey's posthoc test).

These two studies show that inclusion of pomegranate juice in the diet of rodents for 2 to 6 weeks provides substantial protection against oxidative stress-induced damage to neurons exposed to hypoxia and ischemia, possibly by neutralizing oxidants generated during H-I. Our data support the notion that pomegranate polyphenols also enhance the capacity of endogenous brain antioxidant defenses. The effect seen with pomegranate polyphenols in adult H-I is more efficacious than that seen with a comparable diet containing blueberries; the maximum effect seen with blueberries was 57% reduction of H-I damage[61] whereas pomegranates reduced ischemic damage by as much as 93%.

5.7 EFFECTS OF POMEGRANATES ON INFLAMMATION

In our laboratory, male rats that drank 0.0014, 0.014, or 0.14 mg/ml of total pome-granate polyphenols in water for 2, 4, or 6 weeks were euthanized, perfused tran-scardially with saline, and brains were removed. The brain stems were homogenized followed by Western gel electrophoresis to measure concentrations of COX-2 (EC 1.14.99.1). Beta-actin was used to confirm that similar amounts of proteins were loaded onto each lane. Pomegranate juice consumption caused a reduction of the amount of the proinflammatory enzyme COX-2 in brain homogenates while actin content was the same in each gel (Figure 5.4). The effect was greatest after 6 weeks of diet. This suggests that pomegranate polyphenols may have anti-inflammatory effects in the brains of rats. In another study,[62] anthocyanin mixtures from other plant species *(Amelanchier alnifolia, A. arborea,* and *A. canadensis)* inhibited COX-1 and -2 enzymes by 51 to 76%. The positive controls (aspirin, Celebrex, and Vioxx) inhibited COX-2 by 69 to 85%, indicating a comparable effect of anthocy-anins.[62] In another study, the anthocyanins delphinidin and cyanidin inhibited LPS (lipopolysaccharide)-induced COX-2 expression in isolated macrophages, but pel-argonidin, peonidin, and malvidin did not,[63] indicating a selective response. This effect was dose-dependent and likely due to blockade of MAPK-mediated pathways by anthocyanins.[63] Strawberry extracts, rich in ellagic acid much like pomegranates, have been shown to inhibit COX enzymes *in vitro,*[64] which would modulate the inflammatory process. These studies are supportive of our results and suggest that the reductions of COX-2 levels seen with pomegranates are due to the presence of anthocyanins and ellagic acid.

5.8 CONCLUSIONS

Antioxidants in general lower risk of cardiovascular events by inhibition of LDL-oxidation, promotion of atherosclerotic plaque stability, improved vascular endot-helial function, and decreased tendency for thrombosis.[64] Fruits, vegetables, and beverages rich in antioxidant polyphenols have been associated with decreased risk of cardiovascular and other aging-related diseases.[14] The concept that flavonoids and other phenolic compounds are responsible for the health effects of these foods is supported by some *in vitro* studies, animal experiments,[13,14] and clinical studies.[65] Epidemiologic data suggest that beneficial effects on cardiovascular diseases (but not on cancer) are observed in humans fed flavonoids, especially flavonols, flavones, and catechins.[65] The results summarized here have shown that daily ingestion of pomegranate juice, rich in polyphenol antioxidants including flavonoids and ellagic acid,[15,16] has benefits related to cardiovascular disease:

- Slowing the progression and onset of atherosclerosis and dyslipidemia[42,55,56]
- Reducing blood pressure in hypertensive patients[57]
- Reducing blood clotting[60] possibly via slowing of platelet activation
- Neuroprotection from brain ischemia[60] (Sweeney et al., data presented here)

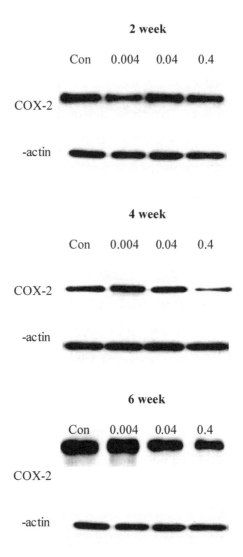

FIGURE 5.4 Effect of drinking pomegranate polyphenols for 2, 4, or 6 weeks, on the levels of cyclooxygenase-2 (COX-2) in rat brains.

- Increased antioxidant status of the brain (Sweeney et al., data presented here)
- Anti-inflammatory effects via reduced COX-2 expression (Sweeney et al., data presented here)

Modifiable risk factors for suffering a first stroke and cerebrovascular disease include hypertension and dyslipidemia.[3,8,10] Therefore, daily pomegranate consumption may play a role in the prevention of the onset of stroke.

Important events during a stroke include atherosclerotic plaque rupture and thrombosis[22] with excessive platelet activation;[23] the resulting hypoxia induces ROS generation (oxidative stress) both acutely and subacutely,[25–31] leading to induction of COX-2 and other inflammatory markers.[24,35–38] Consumption of pomegranate polyphenols has been shown to interfere with all of these events, at least in part. Thus the limited evidence presented here suggests that the presence of pomegranate polyphenols in the brain during a stroke may be beneficial and may dramatically reduce the severity of damage.

Finally, apoptosis contributes to the chronic phase of injury to neurons observed days and months after ischemia,[27,32,33,39,40] partially as a result of ROS generated during reperfusion.[40] Pomegranate feeding to mouse dams was shown to reduce ischemia-induced apoptosis in the offspring.[60] *Ex vivo,* pomegranate polyphenols can quench ROS,[16] and feeding pomegranates elevates the antioxidant capacity of the blood[42–44] and brain (data shown here). Thus, a small body of evidence suggests that continuing a diet of pomegranate juice after a stroke is also beneficial.

There are high mortality and morbidity rates associated with stroke, which cost the U.S. $53.6 billion in 2001.[1] Stroke is a major contributor to long-term disability, cannot be cured, and at present, has no standard effective treatments. The results described in this chapter demonstrate that a modification of the diet to include pomegranate juice may have an impact on the severity and thus financial and emotional costs of stroke.

REFERENCES

1. Heart Disease and Stroke Statistics — 2004 Update. American Heart Association. http://www.americanheart.org/downloadable/heart/1079736729696HDSStats2004UpdateREV3-19-04.pdf
2. Statistics Canada, Causes of Death 2002 (released 2004). http://www.statcan.ca/english/freepub/84-208-XIE/84-208-XIE2004002.htm
3. Heart and Stroke Foundation of Canada: The Growing Burden of Heart Disease and Stroke in Canada 2003. Health Canada, Ottawa, Canada, 2003 (1-896242-30-8, English). http://www.cvdinfobase.ca/cvdbook/CVD_En03.pdf
4. http://www.who.int/violence_injury_prevention/violence/world_report/en/full_en.pdf
5. http://www.benbest.com/lifeext/causes.html#international
6. http://www.who.int/dietphysicalactivity/publications/facts/cvd/en/
7. Ness, A.R. and Powles, J.W., Fruit and vegetables, and cardiovascular disease: a review, *Int. J. Epidemiol.,* 26, 1, 1997.
8. Jenkins, D.J. et al., The Garden of Eden — plant based diets, the genetic drive to conserve cholesterol and its implications for heart disease in the 21st century, *Comp. Biochem. Physiol. Mol. Integr. Physiol.,* 136, 141, 2003.
9. Bazzano, L.A., Serdula, M.K., and Liu, S., Dietary intake of fruits and vegetables and risk of cardiovascular disease, *Curr. Atheroscler. Rep.,* 5, 492, 2003.
10. Srinath Reddy, K. and Katan, M.B., Diet, nutrition and the prevention of hypertension and cardiovascular diseases, *Public Health Nutr.,* 7, 167, 2004.
11. Key, T.J. et al., Mortality in vegetarians and nonvegetarians: detailed findings from a collaborative analysis of 5 prospective studies, *Am. J. Clin. Nutr.,* 70, 516S, 1999.

12. da Luz, P.L. and Coimbra, S.R., Wine, alcohol and atherosclerosis: clinical evidences and mechanisms, *Braz. J. Med. Biol. Res.*, 37, 1275, 2004.

13. Maron, D.J., Flavonoids for reduction of atherosclerotic risk, *Curr. Atheroscler. Rep.*, 6, 73, 2004.

14. Halliwell, B., Rafter, J., and Jenner, A., Health promotion by flavonoids, tocopherols, tocotrienols, and other phenols: direct or indirect effects? Antioxidant or not?, *Amer. J. Clin. Nutr.*, 81, 268S, 2005.

15. Gil, M.I. et al., Antioxidant activity of pomegranate juice and its relationship with phenolic composition and processing, *J. Agric. Food Chem.*, 48, 4581, 2000.

16. Noda, Y. et al., Antioxidant activities of pomegranate fruit extract and its anthocyanidins: delphinidin, cyanidin, and pelargonidin, *J. Agric. Food Chem.*, 50, 166, 2002.

17. Halvorsen, B.L. et al., A systematic screening of total antioxidants in dietary plants, *J. Nutr.*, 132, 461, 2002.

18. Libby, P. and Theroux, P., Pathophysiology of coronary artery disease, *Circ.*, 111, 3481, 2005.

19. National Institute of Neurological Disorders and Stroke (NINDS), Stroke: Hope through Research, NIH Publication No. 99_2222, July 2004. Internet: http://www.ninds.nih.gov/disorders/stroke/detail_stroke.htm#44211105

20. Lakka, H.M. et al., The metabolic syndrome and total and cardiovascular disease mortality in middle-aged men, *J. Amer. Med. Assoc.*, 288, 2709, 2002.

21. Ninomiya, J.K. et al., Association of the metabolic syndrome with history of myocardial infarction and stroke in the Third National Health and Nutrition Examination Survey, *Circ.*, 109, 42, 2003.

22. Ross, R., Atherosclerosis: an inflammatory disease, *N. Engl. J. Med.*, 340, 115, 1999.

23. Cherian, P. et al., Endothelial and platelet activation in acute ischemic stroke and its etiological subtypes, *Stroke*, 34, 2132, 2003.

24. Fagan, S.C. et al., Targets for vascular protection after acute ischemic stroke, *Stroke*, 35, 2220, 2004.

25. Ginsberg, M.D., Adventures in the pathophysiology of brain ischemia: penumbra, gene expression, neuroprotection, *Stroke*, 34, 214, 2003.

26. Ter Horst, G.J. and Korf, J., *Clinical Pharmacology of Cerebral Ischemia*, Humana Press, Totowa, NJ, 1997, chapters 1, 4, 5, 7, 8, 10, and 11.

27. Mergenthaler, P., Dirnagl, U., and Meisel, A., Pathophysiology of stroke: lessons from animal models, *Metab. Brain Dis.*, 19, 151, 2004.

28. del Zoppo, G.J. et al., Polymorphonuclear leucocytes occlude capillaries following middle cerebral artery occlusion and reperfusion in baboons, *Stroke*, 22, 1276, 1991.

29. Prestigiacomo, C.J. et al., CD-18–mediated neutrophil recruitment contributes to the pathogenesis of reperfused but not nonreperfused stroke, *Stroke*, 30, 1110, 1999.

30. Huang, J. et al., Post-ischemic cerebrovascular E-selectin expression mediates tissue injury in murine stroke, *Stroke*, 31, 3047, 2000.

31. Blann, A.D., Ridker, P.M., and Lip, G.Y.H., Inflammation, cell adhesion molecules, and stroke: tools in pathophysiology and epidemiology?, *Stroke*, 33, 2141, 2002.

32. Chan, P.H., Reactive oxygen radicals in signaling and damage in the ischemic brain, *J. Cereb. Blood Flow Metab.*, 21, 2, 2001.

33. Sugawara, T. et al., Neuronal death/survival signaling pathways in cerebral ischemia, *NeuroRx.*, 1, 17, 2004.

34. Lindsberg, P.J. and Grau, A.J., Inflammation and infections as risk factors for ischemic stroke, *Stroke*, 34, 2518, 2003.

35. Zheng, Z. and Yenari, M.A., Post-ischemic inflammation: molecular mechanisms and therapeutic implications, *Neurol. Res.*, 26, 884, 2004.

36. Bemeur, C. et al., Dehydroascorbic acid normalizes several markers of oxidative stress and inflammation in acute hyperglycemic focal cerebral ischemia in the rat, *Neurochem. Int.*, 46, 399, 2005.

37. Gidday, J.M. et al., Leukocyte-derived matrix metalloproteinase-9 mediates blood–brain barrier breakdown and is proinflammatory after transient focal cerebral ischemia, *Am. J. Physiol. Heart Circ. Physiol.*, 289, H558, 2005.

38. Asahi, M. et al., Role for matrix metalloproteinase 9 after focal cerebral ischemia: effects of gene knockout and enzyme inhibition with BB-94, *J. Cereb. Blood Flow Metab.*, 20, 1681, 2000.

39. Li ,Y. et al., Ultrastructural and light microscopic evidence of apoptosis after middle cerebral artery occlusion in the rat, *Am. J. Pathol.*, 146, 1045, 1995.

40. Chan, P.H., Future targets and cascades for neuroprotective strategies, *Stroke*, 35, 2748, 2004.

41. Murthy, K.N.C., Jayaprakasha, G.K., and Singh, R.P., Studies on antioxidant activity of pomegranate *(Punica granatum)* peel extract using *in vivo* models, *J. Agric. Food Chem.*, 50, 4791, 2002.

42. Aviram, M. et al., Pomegranate juice consumption reduces oxidative stress, atherogenic modifications to LDL, and platelet aggregation: studies in humans and in atherosclerotic apolipoprotein E-deficient mice, *Am. J. Clin. Nutr.*, 71, 1062, 2000.

43. Cerdá, B. et al., Repeated oral administration of high doses of the pomegranate ellagitannin punicalagin to rats for 37 days is not toxic, *J. Agric. Food Chem.*, 51, 349, 2003.

44. Seeram, N.P., Lee, R., and Heber, D., Bioavailability of ellagic acid in human plasma after consumption of ellagitannis from pomegranate (*Punica granatum* L.) juice, *Clin. Chim. Acta*, 348, 63, 2004.

45. Kamencic, H. et al., Promoting glutathione synthesis after spinal cord trauma decreases secondary damage and promotes retention of function, *FASEB J.*, 15, 243, 2001.

46. Wu, L. et al., Dietary approach to attenuate oxidative stress, hypertension and inflammation in the cardiovascular system, *Proc. Nat. Acad. Sci.*, 101, 7094, 2004.

47. Schafer, F.Q. and Buettner, G.R., Redox environment of the cell as viewed through the redox state of the glutathione disulfide/glutathione couple, *Free Radic. Biol. Med.*, 30, 1191, 2001.

48. Ben-Yoseph, O., Boxer, P.A., and Ross, B.D., Assessment of the role of the glutathione and pentose phosphate pathways in the protection of primary cerebrocortical cultures from oxidative stress, *J. Neurochem.*, 66, 2329, 1996.

49. Warner, D.S., Sheng, H., and Batinic-Haberle, I., Oxidants, antioxidants and the ischemic brain, *J. Exp. Biol.*, 207, 3221, 2004.

50. Juurlink, B.H., Management of oxidative stress in the CNS: the many roles of glutathione, *Neurotox. Res.*, 1, 119, 1999.

51. Paterson, P.G. and Juurlink, B.H., Nutritional regulation of glutathione in stroke, *Neurotox. Res.*, 1, 99, 1999.

52. Myhrstad, M.C. et al., Flavonoids increase the intracellular glutathione level by transactivation of the γ-glutamylcysteine synthetase catalytical subunit promoter, *Free Radic. Biol. Med.*, 32, 386, 2002.

53. Carlsen, H. et al., Berry intake increases the activity of the γ-glutamylcysteine synthetase promoter in transgenic reporter mice, *J. Nutr.*, 133, 2137, 2003.

54. Cotgreave, I.A. and Gerdes, R.G., Recent trends in glutathione biochemistry. Glutathione-protein interactions: a molecular link between oxidative stress and cell proliferation?, *Biochem. Biophys. Res. Commun.*, 242, 1, 1998.

55. Kaplan, M. et al., Pomegranate juice supplementation to atherosclerotic mice reduces macrophage lipid peroxidation, cellular cholesterol accumulation and development of atherosclerosis, *J. Nutr.*, 131, 2082, 2001.

56. Aviram, M. et al., Pomegranate juice flavonoids inhibit low-density lipoprotein oxidation and cardiovascular diseases: studies in atherosclerotic mice and in humans, *Drugs Exper. Clin. Res.*, 28, 49, 2002.

57. Aviram, M. et al., Pomegranate juice consumption for 3 years by patients with carotid artery stenosis reduces common carotid intima-media thickness, blood pressure and LDL oxidation, *Clin. Nutr.*, 23, 423, 2004.

58. Aviram, M. and Dornfeld, L., Pomegranate juice consumption inhibits serum angiotensin converting enzyme activity and reduces systolic blood pressure, *Atheroscler.*, 158, 195, 2001.

59. Polagruto, J.A. et al., Effects of flavonoid-rich beverages on prostacyclin synthesis in humans and human aortic endothelial cells: association with *ex vivo* platelet function, *J. Med. Food*, 6, 301, 2003.

60. Loren, D.J. et al., Maternal dietary supplementation with pomegranate juice is neuroprotective in an animal model of neonatal hypoxic-ischemic brain injury, *Ped. Res.*, 57, 858, 2005.

61. Sweeney, M.I. et al., Feeding of diets enriched in lowbush blueberries *(Vaccinium angustifolium)* for six weeks decreases stroke severity in rats, *Nutr. Neurosci.*, 5, 427, 2002.

62. Adhikari, D.P. et al., Quantification and characterisation of cyclo-oxygenase and lipid peroxidation inhibitory anthocyanins in fruits of Amelanchier, *Phytochem. Anal.*, 16, 175, 2005.

63. Hou, D.X. et al., Anthocyanidins inhibit cyclooxygenase-2 expression in LPS-evoked macrophages: structure-activity relationship and molecular mechanisms involved, *Biochem. Pharmacol.* 70, 417, 2005.

64. Hannum, S.M., Potential impact of strawberries on human health: a review of the science, *Crit. Rev. Food Sci. Nutr.* 44, 1, 2004.

65. Arts, I.C.W. and Hollman, P.C.H., Polyphenols and disease risk in epidemiologic studies, *Amer. J. Clin. Nutr.*, 81, 317S, 2005.

6 Anticancer Potential of Pomegranate

Shishir Shishodia, Lynn Adams, Indra D. Bhatt, and Bharat B. Aggarwal

CONTENTS

6.1 INTRODUCTION

The pomegranate's medicinal qualities have been known for thousands of years. References in the Bible and Roman mythology mention the tree's unique healing powers, and some Middle Eastern, Asian, and South American people still chew its bark, petals, and peel to treat conditions as diverse as dysentery and diseases of the mouth and gums.[1] Modern research has shown that the pomegranate contains polyphenols and anthocyanidins that are powerful free-radical scavengers and are more effective against disease than are those in red wine and green tea.[2] It is widely used in traditional medicine to cure inflammation, diabetes, cardiac disease, AIDS, ischemia, and cancer (Figure 6.1). On this basis, the possible anticarcinogenic effects of the pomegranate have been further explored. For example, the application of pomegranate extract to the skin of mice before they were exposed to a carcinogenic agent was shown to inhibit the appearance of erythemas and hyperplasia and the activity of epithelial ornithine decarboxylase.[2] The pomegranate has also been shown to induce programmed cell death and to inhibit tumor invasion, proliferation, and angiogenesis. It targets several proteins in the cell-signaling pathway (Figure 6.2). The current review, therefore, will focus on the anticancer potential of the pomegranate and its components.

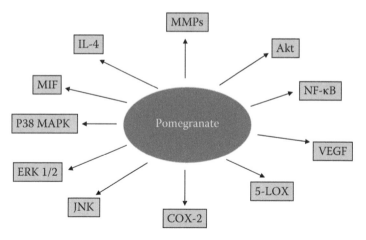

FIGURE 6.1 Disease targets of pomegranate.

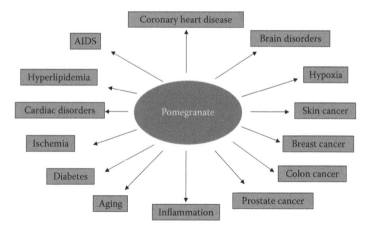

FIGURE 6.2 Molecular targets of pomegranate. (MMPs, matrix metalloproteinases; VEGF, vascular endothelial growth factor; LOX, lipoxygenase; MAPK, mitogen-activated protein kinase; MIF, migration inhibitory factor; JNK, c-Jun N-terminal kinase; ERK1/2, extracellular signal regulated kinase.)

6.2   CHEMICAL COMPONENTS OF THE POMEGRANATE

The unique biochemistry of the pomegranate tree is quite intriguing. In addition to the high levels of antioxidant-rich tannins and flavonoids in its juice and peel, the crushed and dry seeds of its fruit produce a distinct oil, about >60% of which is a very rare 18-carbon fatty acid, also referred to as punicic acid. This oil contains isoflavone genistein, the phytoestrogen coumestrol, and the sex steroid estrone. In fact, the pomegranate tree is one of the only plants in nature known to contain estrone.[3] Its estrone-containing nature may in part explain its therapeutic efficacy, given that several lines of evidence suggest a possible role of phytoestrogens in preventing a range of diseases, not least of which are the hormonally dependent cancers. Online biochemical detection coupled with mass spectrometry has revealed three estrogenic compounds — luteolin, quercetin, and kaempferol — in pomegranates.[3] Other major components of pomegranate juice are ellagic acid, caffeic acid, and punicic acid; all are found in the aqueous or oily compartments of the pomegranate fruit, and each belongs to a different representative chemical class with known anticancer activities.

Anthocyanidins are components of pomegranate fruit that contribute to the antioxidant activity. Anthocyanidins (delphinidin, cyanidin, and pelargonidin) have been shown to have free-radical scavenging activities and inhibitory effects on lipid peroxidation. They inhibit hydrogen peroxide-induced lipid peroxidation in the rat brain homogenates.[4] Evidence suggests that polyphenolic antioxidants contained in pomegranate juice can contribute to the reduction of oxidative stress and atherogenesis through the activation of redox-sensitive genes ELK-1 and p-JUN and increased eNOS expression.[5]

6.3 ANTITUMOR EFFECTS OF THE POMEGRANATE

Various parts of the pomegranate fruit have been shown to exert antiproliferative effects on tumor cells. For example, the polyphenols in the fermented juice of pomegranates have been shown to exert anticancer effects on human breast cancer cells *in vitro*.[3] Mehta et al. further showed that whole pomegranate seed oil is even more chemopreventive against breast cancer than are these polyphenols.[6] In their study of the chemopreventive and adjuvant therapeutic potential of pomegranate components in human breast cancer, Kim et al.[7] found that all the components blocked endogenous active estrogen biosynthesis and aromatase activity by 60 to 80%. The inhibition of cell lines by fermented juice and pericarp polyphenols was highest in estrogen-dependent MCF-7 cells, somewhat lower in estrogen-independent MDA-MB-231 cells, and lowest in normal human breast epithelial MCF-10A cells. In both MCF-7 and MDA-MB-231 cells, fermented pomegranate juice polyphenols consistently showed approximately twice the antiproliferative effect that fresh pomegranate juice polyphenols showed. In addition, pomegranate seed oil effected a 90% inhibition of proliferation of MCF-7 cells at 100 µg/ml medium, a 75% inhibition of invasion of MCF-7 cells across a Matrigel membrane at 10 µg/ml, and a 54% apoptosis in MDA-MB-435 estrogen receptor–negative metastatic cells at 50 µg/ml. Furthermore, in a murine mammary gland organ culture, fermented juice polyphenols effected a 47% inhibition of cancerous lesion formation induced by the carcinogen 7,12-dimethylbenz[*a*]anthracene. These observations support the adjuvant therapeutic application of pomegranate in human breast cancer.[7]

The seed oil, juice, fermented juice, and peel extract of the pomegranate have also been shown to exert suppressive effects on human breast cancer cells *in vitro*.[8] Lansky et al. examined the various dissimilar biochemical fractions originating in anatomically discrete sections of the pomegranate fruit for their synergistic effect against the proliferation, metastatic potential, and phospholipase A2 expression of human prostate cancer cells *in vitro*.[9] They found that these fractions had supra-additive, complementary, and synergistic effects.[9] Similarly, Albrecht et al. examined the effects of pomegranate seed oil, fermented juice polyphenols, and pericarp polyphenols on the growth of human prostate cancer cell xenografts *in vivo* and on the cells' proliferation, cell cycle distribution, apoptosis, gene expression, and invasion across Matrigel *in vitro*. All the three components acutely inhibited the *in vitro* proliferation of LNCaP, PC-3, and DU 145 human prostate cancer cell lines but had much less profound effect on normal prostate epithelial cells. These effects were mediated by changes in both the cell cycle distribution and the induction of apoptosis. In particular, all agents potently suppressed PC-3 invasion through Matrigel and PC-3 xenograft growth in athymic mice.[10]

Pomegranate fruit extract treatment of highly aggressive human prostate cancer PC-3 cells resulted in a dose-dependent inhibition of cell growth/cell viability and induction of apoptosis. The antiproliferative effects of pomegranate fruit extract (PFE) involves the induction of proapoptotic gene products, Bax and Bak and down-regulation of the antiapoptotic genes Bcl-X_L and Bcl-2. PFE treatment also led to induction of WAF1/p21, KIP1/p27, and a decrease in cyclins D1, D2, E; and cyclin-dependent kinase (cdk) 2, cdk4, and cdk6 expression. Oral administration of PFE to athymic nude mice implanted with androgen-sensitive CWR22Rnu1 cells resulted in a significant inhibition in tumor growth concomitant with a significant decrease in serum prostate-specific antigen levels.[11]

Several studies demonstrate that cholesterol accumulates in solid tumors and that cholesterol homeostasis breaks down in the prostate with aging and with the transition to the malignant state.[12] Pomegranate juice has been demonstrated to inhibit the biosynthesis of cholesterol in macrophage cells.[13] These observations suggest that pomegranate juice may have cancer-chemopreventive as well as cancer-chemotherapeutic effects against prostate cancer.

Pomegranate juice, ellagic acid, punicalagin, and total pomegranate tannins have been shown to induce apoptosis in HT-29 colon cells.[14] However, in HCT116 colon cells, apoptosis was induced by ellagic acid, punicalagin, and total pomegranate tannin but not by pomegranate juice.[14] Pomegranate seed oil is composed of more than 70% of the conjugated linolenic acids shown to suppress colon carcinogenesis. Not surprisingly, then, dietary pomegranate seed oil significantly inhibited the incidence of azoxymethane-induced colonic adenocarcinomas.[15]

Excessive human exposure to solar ultraviolet (UV) radiation, particularly to its UV-B component, causes many adverse effects, including erythema, hyperplasia, hyperpigmentation, immunosuppression, photoaging, and skin cancer, that the pomegranate can counter. In this regard, the anthocyanidins (such as delphinidin, cyanidin, and pelargonidin) and hydrolyzable tannins (such as punicalin, pedunculagin, punicalagin, and gallagic and ellagic acid esters of glucose) in the pomegranate possess strong antioxidant and anti-inflammatory properties that protect from ultraviolet radiation. In addition, the pomegranate fruit extract (PFE) has been shown to inhibit UV-B–mediated phosphorylation of mitogen-activated protein kinase and nuclear factor (NF)-κB activation.[16] The dermal application of PFE also suppressed 12-O-tetradecanoylphorbol-13-acetate (TPA)-induced skin tumor promotion in 7,12-dimethylbenz(*a*)anthracene–initiated CD-1 mice.[17] Compared with mice that did not receive PFE, the animals pretreated with the extract showed a substantially reduced tumor incidence and a lower total body tumor burden when assessed according to the total number of tumors per group, the percentage of mice with tumors, and the number of tumors per animal.

Pomegranate seed oil has also been shown to significantly decrease tumor incidence, tumor multiplicity, and TPA-induced ornithine decarboxylase activity in a two-stage skin carcinogenesis model in mice. These results highlight the potential of pomegranate as a safe and effective chemopreventive agent against skin cancer.[17]

Flavonoid-rich polyphenol fractions from the pomegranate fruit also exert anti-proliferative, anti-invasive, antieicosanoid, and proapoptotic effects on breast and prostate cancer cells and have antiangiogenic activities *in vitro* and *in vivo*. For example, the flavonoid-rich fractions from pomegranate juice were shown to be potential differentiation-promoting agents of human HL-60 promyelocytic leukemia cells.[18] In addition, the effect of the boiled extract of pomegranate peel was examined in the human cell lines Raji and P3HR-1. The proliferation and viability of these tumor cells were dose-dependently reduced by the pomegranate extract. Collectively, these studies confirm the potent antitumor effects of the various components of the pomegranate fruit.[8]

6.4 EFFECT ON NF-κB ACTIVATION

The various components of the pomegranate have been tested for their effects on NF-κB. NF-κB is a transcription factor that is activated in response to various stimuli, including cytokines, mitogens, carcinogens, chemotherapeutic agents, endotoxin, physical and chemical stresses, radiation, hypoxia, and other inflammatory stimuli. Furthermore, constitutively active NF-κB is frequently encountered in a wide variety of tumors. NF-κB activation has been shown to regulate the expression of over 200 genes involved in cellular transformation, proliferation, antiapoptosis, angiogenesis, invasion, and metastasis. Whereas most carcinogens and tumor promoters activate NF-κB, chemopreventive agents can suppress this activation. For example, the activation of NF-κB in vascular endothelial cells in response to biochemical or biomechanical stimuli is associated with vascular pathologies such as atherosclerosis; however, pomegranate wine has been shown to inhibit tumor necrosis factor α-(TNF) or shear stress-mediated NF-κB activation in vascular endothelial cells.[19]

Afaq and colleagues examined the effect of PFE on UV-B–induced adverse effects in normal human epidermal keratinocytes (NHEK). The treatment of NHEK with PFE (10 to 40 μg/mL) for 24 h before UV-B (40 mJ/cm^2) exposure inhibited UV-B–mediated phosphorylation of extracellular signal regulated kinase (ERK)l/2, c-jun N terminal kinase (JNK) 1/2, and p38 mitogen activated protein kinase in a dose-dependent manner. PFE treatment of NHEK also resulted in a dose- and time-dependent inhibition of UV-B–mediated degradation and phosphorylation of IκBα and in activation of IKKα. It was also shown that PFE treatment of NHEK resulted in a dose- and time-dependent inhibition of UV-B–mediated nuclear translocation and phosphorylation of NF-κB/p65 at Ser(536).[16,20] Additionally, in the HT-29 human colon cancer cell line, PFE was shown to reduce TNFα-induced phosphorylation of the p65 subunit and binding to the NF-κB response element 6.4-fold. TPT suppressed

NF-κB binding 10-fold, punicalagin 3.6-fold where ellagic acid (another major pomegranate component) was ineffective. In this same study, PFE also abolished TNFα-induced activation of AKT, a kinase involved in the induction of NF-κB activity (unpublished data). Moreover, in our study of the effects of the various pomegranate components on NF-κB and AKT activation, we found that pomegranate juice, total pomegranate tannins, and punicalagin suppressed NF-κB and AKT activation, leading to inhibition of cyclooxygenase-2 (COX-2) expression and of cellular proliferation in colon cancer cells (unpublished data). Therefore, the anti-inflammatory properties of polyphenolic phytochemicals in the pomegranate can play an important role in the modulation of inflammatory cell survival signaling.

6.5 EFFECT ON COX-2

COX-2 is an enzyme induced by inflammatory and mitogenic stimuli and causes enhanced synthesis of prostaglandins in inflamed and neoplastic tissues. It is associated with cell proliferation and growth in various cancerous conditions and is a regulatory factor for several pathways that can result in cancer. Additionally, COX-2 makes cells resistant to apoptosis and promotes angiogenesis, metastasis, and the cancer cell cycle by controlling several targets.

Afaq et al. found that the topical application of PFE (2 mg/mouse) 30 min before TPA application to mouse skin (3.2 nmol/mouse) inhibited the TPA-mediated increase in the protein expression of COX-2 in a time-dependent manner.[20] Likewise, Schubert et al. found in their examination of the antioxidant and eicosanoid enzyme inhibitory properties of pomegranate fermented juice and seed oil flavonoids that pomegranate fermented juice and cold-pressed seed oil showed strong antioxidant activity, which was close to that of butylated hydroxyanisole and green tea and significantly greater than that of red wine. In fact, the flavonoids extracted from cold-pressed seed oil showed a 31 to 44% inhibition of sheep COX and a 69 to 81% inhibition of soybean lipoxygenase. Additionally, the flavonoids extracted from pomegranate fruit juice showed a 21 to 30% inhibition of soybean lipoxygenase, although no significant inhibition of sheep COX was observed.[19] In addition, Adams et al. examined the effects of pomegranate juice extract on inflammatory cell signaling proteins in the HT-29 human colon cancer cell line. This study found that PFE significantly suppressed TNFα-induced COX-2 protein expression by 79%, total pomegranate tannin extract 55%, and punicalagin 48%.[21] Thus, pomegranate extracts could be potential anti-inflammatory agents because of their COX-2 inhibitory activities.

6.6 EFFECT ON ANGIOGENESIS

It is now well recognized that the induction of the tumor vasculature or angiogenesis is critical for the progression of tumors. This tumor vascularization has been shown to be dependent on the chemokines (e.g., monocyte chemoattractant protein-1, interleukin [IL]-8) and growth factors (e.g., tumor necrosis factor, vascular endothelial growth factor) produced by macrophages, neutrophils, and other inflammatory cells.[22] The production of these angiogenic factors has been shown to be regulated by NF-κB activation.[23]

Toi et al. evaluated the antiangiogenic potential of pomegranate seed oil or fermented juice polyphenols on angiogenic regulation by measuring the levels of vascular endothelial growth factor, IL-4, and migration inhibitory factor in the conditioned media of estrogen-sensitive (MCF-7) and estrogen-resistant (MDA-MB-231) human breast cancer cells and of immortalized normal human breast epithelial cells (MCF-10A).[24] Their results showed a significant potential for down-regulation of angiogenesis by pomegranate fractions. Toi et al. also demonstrated an antiproliferative effect on angiogenesis in human umbilical vein endothelial cells and in myometrial and amniotic fluid fibroblasts and an inhibition of tubule formation in human umbilical vein endothelial cells in an *in vitro* model that used glass carrier beads. The use of pomegranate fractions also led to a significant decrease in new blood vessel formation in a chicken chorioallantoic membrane model *in vivo*. These observations suggest that the antiangiogenic effects of pomegranate components are most likely mediated by the suppression of NF-κB.

6.7 EFFECT ON INVASION

When cancer treatment is ineffective, the remaining tumor cells inevitably infiltrate the surrounding normal tissue, which leads to tumor recurrence. Recent studies have shown that the ability of tumor cells to digest the extracellular matrix by secreting proteolytic enzymes correlates well with their degree of tissue invasiveness. For most primary human tumors, invasion is thought to be accomplished, at least in part, by proteases — namely, serine, cysteine, and metalloproteinases — that penetrate connective-tissue barriers, induce vascular remodeling, and destroy normal tissue. Several proteases, including matrix metalloproteinases and the serine protease urokinase-type plasminogen activator, that influence the invasive characteristics of tumors are regulated by NF-κB.[25–27]

Pomegranate extract has been shown to inhibit tumor invasion. In this regard, pomegranate hampered IL-1β–induced expression of matrix metalloproteinases by inhibiting the activation of mitogen-activated protein kinases and NF-κB in human chondrocytes *in vitro*.[28] In another study, various components of pomegranate, including ellagic acid, caffeic acid, luteolin, and punicic acid, were tested as potential inhibitors of the *in vitro* invasion of human PC-3 prostate cancer cells across Matrigel artificial membranes. Although all compounds significantly inhibited invasion when administered individually, a supra-additive inhibition occurred when the components were administered together.[29] Albrecht et al. similarly examined the effects of pomegranate oil and polyphenols and of pericarp polyphenols on human prostate cancer cell invasion across Matrigel *in vitro*. Again, all agents potently suppressed PC-3 invasion across Matrigel. Overall, these observations confirm the significant anti-invasive activity of pomegranate-derived materials against human prostate cancer.[10]

6.8 CONCLUSION

As previously stated, the medicinal use of pomegranate to treat various ailments has been well described in ancient texts and mythology. In more recent times, the scientific basis supporting this use has now become clear, with modern research providing an explanation of the numerous medicinal effects of the pomegranate. Specifically, the pomegranate and its various components have been shown to target a number of pathways that are responsible for various diseases, including cancers, and hence to be effective against many different disorders. In particular, pomegranate extracts have shown enough potential against tumors in numerous preclinical studies to warrant clinical trials.

Although modern drug design emphasizes the development of single agents with specific targets, the fact that whole pomegranate extract has been shown to be more efficacious than its individual components (a concept known as herbal synergy) suggests the limitations of this approach. Thus, the time to develop a dietary agent or drug that consists of a mixture of distinct molecules has come, however huge the challenge may be scientifically. Therefore, not only is an in-depth study to define the active agent(s) in PFE responsible for its antitumor effects warranted, additional clinical trials to further assess the chemopreventive and adjuvant therapeutic applications of the pomegranate in human cancers are also appropriate.

ACKNOWLEDGMENT

Supported by funds from the Clayton Foundation for Research (to BBA), a Department of Defense U.S. Army Breast Cancer Research Program grant (BC010610, to BBA), a P01 grant (CA-91844) on lung cancer chemoprevention from the National Institutes of Health (to BBA), a P50 Head and Neck Cancer SPORE grant from the National Institutes of Health (to BBA), and Cancer Center Core Grant CA-16672 (to BBA).

REFERENCES

1. Longtin, R., The pomegranate: nature's power fruit? *J. Natl. Cancer Inst.,* 95, 346, 2003.
2. Burton, A., Chemoprevention: eat ginger, rub on pomegranate, *Lancet Oncol.,* 4, 715, 2003.
3. van Elswijk, D.A. et al., Rapid dereplication of estrogenic compounds in pomegranate *(Punica granatum)* using on-line biochemical detection coupled to mass spectrometry, *Phytochemistry,* 65, 233, 2004.
4. Noda, Y. et al., Antioxidant activities of pomegranate fruit extract and its anthocyanidins: delphinidin, cyanidin, and pelargonidin, *J. Agric. Food Chem.,* 50, 166, 2002.
5. de Nigris, F. et al., Beneficial effects of pomegranate juice on oxidation-sensitive genes and endothelial nitric oxide synthase activity at sites of perturbed shear stress, *Proc. Natl. Acad. Sci. USA,* 102, 4896, 2005.
6. Mehta, R. and Lansky, E.P., Breast cancer chemopreventive properties of pomegranate *(Punica granatum)* fruit extracts in a mouse mammary organ culture, *Eur. J. Cancer Prev.,* 13, 345, 2004.
7. Kim, N.D, et al., Chemopreventive and adjuvant therapeutic potential of pomegranate *(Punica granatum)* for human breast cancer, *Breast Cancer Res. Treat.,* 71, 203, 2002.

8. Settheetham, W. and Ishida, T., Study of genotoxic effects of antidiarrheal medicinal herbs on human cells *in vitro, Southeast Asian J. Trop. Med. Public Health,* 26 Suppl 1, 306, 1995.

9. Lansky, E.P. et al., Possible synergistic prostate cancer suppression by anatomically discrete pomegranate fractions, *Invest. New Drugs,* 23, 11, 2005.

10. Albrecht, M. et al., Pomegranate extracts potently suppress proliferation, xenograft growth, and invasion of human prostate cancer cells, *J. Med. Food.,* 7, 274, 2004.

11. Malik, A. et al., Pomegranate fruit juice for chemoprevention and chemotherapy of prostate cancer, *Proc. Natl. Acad. Sci. USA,* 102, 14813, 2005.

12. Freeman, M.R. and Solomon, K.R., Cholesterol and prostate cancer, *J. Cell Biochem.,* 91, 54, 2004.

13. Fuhrman, B., Volkova, N., and Aviram, M., Pomegranate juice inhibits oxidized LDL uptake and cholesterol biosynthesis in macrophages, *J. Nutr. Biochem.,* 16, 570, 2005.

14. Seeram, N.P. et al., *In vitro* antiproliferative, apoptotic and antioxidant activities of punicalagin, ellagic acid and a total pomegranate tannin extract are enhanced in combination with other polyphenols as found in pomegranate juice, *J. Nutr. Biochem.,* 16, 360, 2005.

15. Kohno, H. et al., Pomegranate seed oil rich in conjugated linolenic acid suppresses chemically induced colon carcinogenesis in rats, *Cancer Sci.,* 95, 481, 2004.

16. Afaq, F. et al., Pomegranate fruit extract modulates UV-B-mediated phosphorylation of mitogen-activated protein kinases and activation of nuclear factor kappa B in normal human epidermal keratinocytes paragraph sign, *Photochem. Photobiol.,* 81, 38, 2005.

17. Hora, J.J. et al., Chemopreventive effects of pomegranate seed oil on skin tumor development in CD1 mice, *J. Med. Food,* 6, 157, 2003.

18. Kawaii, S. and Lansky, E.P., Differentiation-promoting activity of pomegranate *(Punica granatum)* fruit extracts in HL-60 human promyelocytic leukemia cells, *J. Med. Food,* 7, 13, 2004.

19. Schubert, S.Y., Lansky, E.P., and Neeman, I., Antioxidant and eicosanoid enzyme inhibition properties of pomegranate seed oil and fermented juice flavonoids, *J. Ethnopharmacol.,* 66, 11, 1999.

20. Afaq, F. et al., Anthocyanin- and hydrolyzable tannin-rich pomegranate fruit extract modulates MAPK and NF-kappaB pathways and inhibits skin tumorigenesis in CD-1 mice, *Int. J. Cancer,* 113, 423, 2005.

21. Adams, L. et al., Pomegranate juice, total pomegranate ellagitanins, and punicalagin suppress inflammatory cell signaling in colon cancer cells. *J. Agric. Food Chem.,* 54, 980, 2006.

22. Loch, T. et al., Vascular endothelial growth factor (VEGF) and its role in neoplastic processes, *Postepy Hig. Med. Dosw.,* 55, 257, 2001.

23. Chilov, D. et al., Genomic organization of human and mouse genes for vascular endothelial growth factor C, *J. Biol. Chem.,* 272, 25176, 1997.

24. Toi, M. et al., Preliminary studies on the anti-angiogenic potential of pomegranate fractions *in vitro* and *in vivo, Angiogenesis,* 6, 121, 2003.

25. Farina, A.R. et al., Transcriptional up-regulation of matrix metalloproteinase-9 expression during spontaneous epithelial to neuroblast phenotype conversion by SK-N-SH neuroblastoma cells, involved in enhanced invasivity, depends upon GT-box and nuclear factor kappaB elements, *Cell Growth Differ.,* 10, 353, 1999.

26. Bond, M. et al., Synergistic upregulation of metalloproteinase-9 by growth factors and inflammatory cytokines: an absolute requirement for transcription factor NF-kappa B, *FEBS Lett.,* 435, 29, 1998.

27. Novak, U., Cocks, B.G., and Hamilton, J.A., A labile repressor acts through the NFkB-like binding sites of the human urokinase gene, *Nucleic Acids Res.,* 19, 3389, 1991.
28. Ahmed, S. et al., *Punica granatum* L. extract inhibits IL-1{beta}-induced expression of matrix metalloproteinases by inhibiting the activation of MAP kinases and NF-{kappa}B in human chondrocytes *in vitro, J. Nutr.,* 135, 2096, 2005.
29. Lansky, E.P, et al., Pomegranate *(Punica granatum)* pure chemicals show possible synergistic inhibition of human PC-3 prostate cancer cell invasion across Matrigel, *Invest New Drugs,* 23, 121, 2005.

7 Molecular Mechanisms of Chemoprevention of Cancer by Pomegranate

Deeba Syed, Arshi Malik, Naghma Hadi, Caroline Schmitt, Farrukh Afaq, and Hasan Mukhtar

CONTENTS

7.1   INTRODUCTION

Despite advances in early detection and improvement in treatment options, cancer continues to be a major health issue in terms of morbidity and mortality. In the U.S. alone, a total of 1,372,910 new cancer cases and 570,280 deaths are expected in the year 2005. Among men, prostate, bronchoalveolar, and colorectal cancers account for more than 56% of all newly diagnosed cases. The three most commonly diagnosed cancers among women are breast, bronchoalveolar, and colorectal, accounting for approximately 55% of estimated cancer cases.[1] Cancer is an aggressive disease that when not detected at an early stage can metastasize to other organs of the body, potentially making even systemic chemotherapy ineffective. Cancer development, termed *carcinogenesis,* involves the evolution of a clinically detectable tumor, which generally takes many years in humans. The process occurs through three well-defined stages known as initiation, promotion, and progression.

Because treatment options for advanced metastasized cancers remain inadequate, developing effective approaches for the prevention of cancer appears to be both a practical and promising strategy to reduce cancer burden. In recent years, the concept of chemoprevention, often referred to as "prevention by delay," is considered a realistic option to achieve this goal. Chemoprevention is a means of cancer control defined by the administration of synthetic or naturally occurring agents to suppress, reverse, or prolong the process of carcinogenesis.[2] It is becoming increasingly clear that chemopreventive compounds in the diet offer great potential in the fight against cancer by inhibiting the process of carcinogenesis through regulation of cell-defensive and cell-death machineries. Recent data from our laboratory and others suggest that agents derived from the fruit pomegranate could function as useful chemopreventive substances against human cancers. Pomegranate has been used in folk medicine to treat inflammation, sore throats, and rheumatism for centuries. Also noted for its antiviral, antihelminthic, antifungal, and antibacterial activities, pomegranate has been cultivated since ancient times throughout the Mediterranean region. About a decade ago, Western scientists became interested in the unique biochemistry of the pomegranate tree as they discovered the high levels of antioxidant-rich tannins and flavonoids in its juice and peel.[3] Several studies have since been conducted to evaluate the efficacy of this naturally occurring botanical antioxidant as an antiproliferative, anti-invasive, and pro-apoptotic agent in various cancer cell lines and animal models. This chapter summarizes the current knowledge, obtained from cell culture and animal models, regarding the chemoprotective and chemotherapeutic potentials of pomegranate in treating cancers of the skin, prostate, and lung.

7.2   CANCER CHEMOPREVENTION

Chemoprevention is a rapidly growing area of preventive oncology that focuses on the prevention of cancer through the use of natural and synthetic compounds. The foreseeable goal of chemoprevention is to slow down the process of carcinogenesis in high-risk individuals who are otherwise cancer free. The logic behind this thinking is that an effective agent can inhibit the onset or slow the advancement of a normal cell to an invasive malignant cell. It is hoped that nontoxic agents could be administered to otherwise healthy individuals who may be at increased risk of cancer to facilitate intervention in the early stages, before the invasive disease begins. There is considerable evidence that once the target population has been identified, adoption of preventive strategies can effectively reduce the risk of major human cancers including breast, lung, prostate, colorectal, ovarian, and bladder cancers. For a variety of reasons, the most important of which are regulatory approval and social acceptance, there is greater emphasis on food- and beverage-derived products as potential cancer chemopreventive substances for human use. This notion is consistent with the reality of life that dictates that in this industrialized world one cannot control the environmental carcinogens and genetic predisposition but can certainly make choices for the food and beverages one consumes. Studies suggest that about 35% of human cancers can be attributed to dietary factors[2] but there is accumulating evidence that regular consumption of fruits and vegetables decreases the risk of cancer. The chemopreventive effects of vegetables and fruits may be attributed to a combined effect of various phenolic phytochemicals that are generally antioxidant in nature, along with vitamins, dietary fibers, indoles, allium compounds, and selenium.[4] Several campaigns have been launched in recent years by the World Health

Organization, the National Cancer Institute, and other organizations to increase the daily food and vegetable intake by the population. A number of human intervention trials are underway to test the ability of the micro- and macronutrients present in these foods to prolong the process of carcinogenesis.

7.3 MOLECULAR MECHANISMS OF CHEMOPREVENTION

Carcinogenesis is a multistage process, strongly influenced by a number of variables such as age, dietary habits, and hormonal balance. The very first phase of carcinogenesis, called *initiation,* is a rapid and irreversible process in which exposure to a carcinogen with consequent covalent interaction of reactive species with target-cell DNA leads to genotoxic damage.[5] Tumor *promotion,* in contrast, is a relatively lengthy and reversible process and is the next phase in which the initiated cells clonally expand by active proliferation resulting in the accumulation of preneoplastic cells. This is followed by the final stage of the disease, the *progression* phase, which involves the growth of the tumor with invasive and metastatic potential.[6]

The ability of chemopreventive agents to retard or halt tumor development is the result of a combination of multiple intracellular responses rather than a single event. Many of these are associated with alterations in cell signaling pathways that regulate cell proliferation and differentiation. Disruption or dysregulation of these pathways may result in malignant transformation of the cells. An ideal chemopreventive agent will be one that is acceptable to the human population and has the ability to restore all dysregulated cellular and molecular pathways of multistage carcinogenesis.[7] This could be achieved through the use of multiple agents, which is a complex undertaking. Alternatively, it might be realized through the application of a single dietary agent capable of interfering at multiple pathways in the carcinogenic process, a very unlikely possibility. Several *in vitro* and *in vivo* studies using pomegranate fruit extract suggest that it can be an effective chemopreventive agent capable of interfering with multiple pathways critically involved in different stages of the development and progression of tumors.

7.4 CHEMICAL CONSTITUENTS OF POMEGRANATE

Pomegranate (*Punica granatum,* Punicaceae), native to semitropical Asia and naturalized in the Mediterranean region, is now cultivated in Afghanistan, India, China, Japan, Russia, and the U.S., particularly in Arizona and California. Edible parts of the pomegranate fruit (about 80% of total fruit weight) are comprised of 80% juice and 20% seed. The layers of seeds are separated by paper-thin white membranes that are bitter to the tongue. The inner membranes and rind are not generally eaten due to their high tannic acid content, but are useful as skin wash. The fruit itself is a rich source of two types of polyphenolic compounds: anthocyanins (such as delphinidin, cyanidin, and pelargonidin), which give the fruit and juice its red color, and hydrolyzable tannins (such as punicalin, pedunculagin, punicalagin, and gallagic and ellagic acid esters of glucose), which account for 92% of the antioxidant activity of the whole fruit.[8] Pomegranate seeds are rich in sugars, polyunsaturated (n-3) fatty acids, vitamins, polysaccharides, polyphenols, and minerals and have high antioxidant activity. When crushed and dried, the seeds produce an oil with 80% punicic acid, the 18-carbon fatty acid, along with the isoflavone genistein, the phytoestrogen coumestrol, and the sex steroid estrone. The seed coat of the fruit contains delphinidin-3-glucoside, delphinidin-3,5-diglucoside, cyanidin-3-glucoside, cyanidin-3,5-diglucoside, pelargonidin-3-glucoside, and pelargonidin-3,5-diglucoside, with delphinidin-3,5-diglucoside being the major anthocyanin in pomegranate juice.[9] Anthocyanidins differ structurally from other flavonoids except flavan-3-ol, and with no carboxyl group in the C-ring are capable of preventing lipid peroxidation of cell or liposome membranes. Delphinidin, with a hydroxyl group at 3′, 4′, and 5′ and cyanidin at 3′ and 4′ positions in the B-ring, can chelate iron and other metals like copper, tin, and aluminum. However, pelargonidin, which has no orthodihydroxy substitution, has been reported to have no chelating capacity with metals.[9] All three anthocyanins were shown to inhibit H_2O_2-induced lipid peroxidation in rat brain homogenates, with delphindin being the most potent and pelargonidin the least.[9] The antimutagenic effect of the bioactive pomegranate compounds has been demonstrated by a decrease in the frequency of genotoxicant-induced chromosomal aberrations in bone marrow cells of mice and rats.[10] Various fractions have been extracted from the pomegranate peel and seeds using ethyl acetate, methanol, and water. These were screened for their antioxidant potential in *in vitro* models, such as beta-carotene-linoleate and 1,1-diphenyl-2-picryl hydrazyl systems.[11] Extraction with methanol was reported to give a higher yield with greater antioxidant activity. The peel exhibited higher activity as compared to seeds, ascribed to its phenolic composition.[11] Another study suggested that pomegranate compared to apple juice is more effective in improving the antioxidant system of aged Wistar rats.[12] The flavonoid-rich polyphenolic antioxidants present in the juice have been shown to contribute to reduction of oxidative stress and atherogenesis with anti-proliferative, anti-invasive, anti-eicosanoid, and pro-apoptotic actions in breast and prostate cancer cells *in vitro* and *in vivo*.[13] Studies have shown that the antioxidant capacity of pomegranate juice is three times that of the popular antioxidant-containing beverages such as red wine and green tea, presumably due to the presence of hydrolyzable tannins in the rind, along with anthocyanins and ellagic acid derivatives.[14]

7.5 POMEGRANATE AND SKIN CANCER

The expanded definition of chemoprevention also encompasses chemotherapy of precancerous lesions, which are called pre-invasive neoplasia, dysplasia, or intraepithelial neoplasia, depending on the organ system. We evaluated the chemopreventive potential of pomegranate using a two-stage skin tumorigenesis protocol in a mouse model. The edible part of pomegranate was extracted with acetone and the product obtained, termed pomegranate fruit extract (PFE), was analyzed with matrix-assisted laser desorption/ionization time of flight mass spectrometry (MALDI-TOF MS).[8] It was found to contain anthocyanins (pelargonidin 3-glucoside, cyanidin

3-glucoside, delphinidin 3-glucoside, pelargonidin 3,5-diglucoside, cyanidin 3,5-diglucoside, and delphinidin 3,5-diglucoside), various ellagitannins, and hydrolyzable tannins.[8] Skin cancer was initiated in CD-1 mice with topical application of the initiating agent, 7,12-dimethylbenzanthracene (DMBA), followed by biweekly applications of the promoting agent, 12-O-tetradecanoylphorbol 13-acetate (TPA). Using this protocol PFE-pretreated mice developed fewer tumors with significant prolongation of the latency period. Animals that did not receive the extract had marked DMBA-initiated, TPA-induced tumor promotion activity, with 100% of the mice developing tumors at 16 weeks whereas at this time only 30% of the treated mice exhibited tumors.[8]

Topical application of TPA to mouse skin is known to result in a number of biochemical alterations, changes in cellular functions, and histological changes leading to skin tumor promotion.[8] Ornithine Decarboxylase (ODC), which forms putrescine from ornithine, is the rate-limiting enzyme in polyamine biosynthesis.[15] Elevated ODC activity and expanded pools of the polyamines are commonly associated with tumorigenesis, and a role of an oncogene-like protein has been assigned to ODC.[16] Agents that block the induction of ODC can prevent tumor formation, and its inhibition has been shown to be a promising tool for screening inhibitors of tumorigenesis.[15] PFE applied to mice skin afforded significant inhibition, in a time-dependent manner against TPA-mediated increase in skin edema and hyperplasia.[8] In addition, significant inhibition of TPA-induced epidermal ODC activity was seen in the treated mice. Cyclooxygenase (COX), another enzyme involved in the mediation of inflammatory process, catalyzes the rate-limiting step in the synthesis of prostaglandins from arachidonic acid.[17,18] Of its two isoforms, COX-1 is constitutively expressed in most tissues and appears to be responsible for maintaining normal physiological functions whereas COX-2 has been shown to be involved in cutaneous inflammation, cell proliferation, and skin tumor promotion.[19] There is compelling evidence that inhibition of COX-2 activity is important not only for alleviating inflammation, but also for the prevention of cancer. TPA application to mice skin resulted in the induction of COX-2, as expected. Interestingly, pretreatment of mice with PFE resulted in decreased expression of COX-2, suggesting that the primary effect of the compound may be against inflammatory responses, with consequent inhibition of tumor promotion.[8]

Mitogen-activated protein kinases (MAPKs) constitute a family of proteins including extracellular signal-regulated kinases (ERKs), stress-activated protein kinases (SAPK/JNKs), and p38 and have a well-documented role in tumor cell proliferation.[8,20] Studies have shown that ERK1/2 and p38 are involved in the transcriptional activation of nuclear factor kappa B (NFκB), which has emerged as one of the most promising molecular targets in the prevention of cancer.[21] NFκB involved in inflammatory and antiapoptotic processes resides in the cytoplasm in an inactive state. Upon activation, inhibitor of IκB kinase (IKK) phosphorylates IκB and is then ubiquitinated and degraded, releasing the NFκB p50-p65 dimer to translocate to the nucleus and activate gene expression.[22] Pretreatment of mice skin with PFE was found to inhibit TPA-mediated phosphorylation of MAPKs as well as NFκB and

IKK activation and phosphorylation and degradation of IκB protein.[8] Inhibition of these events by pomegranate extract may be responsible for its demonstrated anti-tumor-promoting effects.

The seed oil from pomegranate was investigated for its chemopreventive efficacy in another *in vivo* study by Hora et al.[23] In this study, the experimental group of animals was pretreated with 5% pomegranate seed oil prior to TPA application. In mice treated with pomegranate seed oil, both tumor incidence and tumor multiplicity were found to be greatly reduced in treated vs. the control group. After 20 weeks of promotion by DMBA, the average number of tumors per mouse in the untreated group was 20.8 compared to 16.3 in the pomegranate-treated group. The 17% reduction of ODC activity seen in the treated group further demonstrated the potential of pomegranate as a safe and effective chemopreventive agent against skin cancer in the mouse model.[23]

Exploitation of pomegranate extract as a topical dermatological formulation was suggested in one study where methanolic extract of dried pomegranate peel was used to treat excision wounds on the skin of Wistar rats. Treated animals showed good healing compared with controls, with increased hydroxyproline content.[24]

Solar ultraviolet (UV) radiation, depending upon the wavelength, is divided into UVC (200–280 nm), UVB (280–320 nm), and UVA (320–400 nm), of which UVC is effectively filtered out by the ozone layer. It produces a variety of adverse effects that includes sunburns, photoaging, immunosuppression, angiogenesis, photoderma-toses, DNA mutations, and photocarcinogenesis. The endogenous antioxidant capacity of the skin is a major determinant in its response to UV-induced oxidative stress-mediated skin damage. The oxidant/antioxidant imbalance induced by UV results in the generation of reactive oxygen species (ROS) that in turn causes cellular damage by reacting with lipids, proteins, and DNA. In recent years, considerable attention has been focused on the use of naturally occurring botanicals for their potential preventive effect against UV radiation-mediated damage referred to as "photochemopreventive effects."[25] Botanical agents, which generally contain agents that are antioxidant in nature, are being widely used in skin care products and even in customized diets and beverages.[25]

Using normal human epidermal keratinocytes as a test system, we recently showed the remarkable photochemopreventive effects of PFE against UVA[26] and UVB radiation.[27] Our study focused on UV-induced intracellular-signaling cascades in an effort to understand the molecular targets responsible for PFE effects. Different components of UV radiation differ in their biological impact; UVA causes indirect oxidative damage to the cells and exerts its carcinogenic effect due to ROS gener-ation, primarily singlet oxygen.[28] UVB radiation on the other hand directly or indirectly targets the DNA by producing cyclobutane pyrimidine dimers, pyrimi-dine–pyrimidone (6-4) photoproducts, causing oxidative DNA base damage, and generating ROS, oxidative stress, and consequent alterations in a variety of signaling events.[29] Various signal transduction pathways, transcription factors, and changes in gene expression are involved in UV-mediated proliferation, differentiation, or apop-tosis of skin cells.[28,29]

The STATs (signal transducer and activator of transcription) are latent cytoplas-mic transcription factors that become activated in response to extracellular signaling proteins, including growth factors, cytokines, hormones, and peptides. Activation of STAT molecules involves their tyrosine phosphorylation by several tyrosine kinases. Activated STAT then translocates into the nucleus and drives the transcription of specific genes through interactions with DNA elements in the promoters of target genes.[30] A requirement for STAT3 for the development of skin tumors and the maintenance of their autonomous growth has been shown and, more importantly, it was found that mice deficient in STAT3 were completely resistant to skin tumor development.[31] Pomegranate extract pretreatment to human epidermal cells exposed to UVA was found to inhibit the phosphorylation of STAT3 at Tyr[705]. This is significant as STAT3 is thought to be phosphorylated following integration of signals from cell-signaling molecules including ERK and AKT1 induced concurrently by UVA stimulation.

AKT protein kinase, activated by growth factors, is primarily involved in cell survival. Its activation may result in increased proliferation of the epidermal keratinocytes and the ability to resist terminal differentiation during multistage skin carcinogenesis. Pomegranate extract treatment to human epidermal cells results in inhibition of UVA-induced phosphorylation of AKT1 at Ser[473] in a dose-dependent manner.[26] This is important as activation of AKT1 in response to UV-induced oxidant injury may influence cell survival, with a role in carcinogenesis.

MAPK cascade, the three-kinase module conserved in all eukaryotes, has been shown to be activated in response to UVA in various cell lines.[32] MAPK activation may result in cellular proliferation by promoting AP-1 activation and subsequent up-regulation of proto-oncogenes like cyclin D1.[33] It has been reported that PKC-mediated ERK activation by UVA provides a signal for the keratinocytes to escape apoptosis.[34] UVA exposure to normal human epidermal keratinocytes (NHEK) has been shown to result in phosphorylation of ERK1/2 and pretreatment of these cells with pomegranate extract results in a dose-dependent inhibition of UVA-mediated phosphorylation of ERK1/2.

Mammalian target of rapamycin (mTOR) is a serine/threonine kinase involved in the regulation of cell growth through initiation of gene translation. The mTOR kinase initiates translation by activating the p70S6 kinase that in turn phosphorylates the S6 protein of the 40S ribosomal subunit directing the cell toward increased production of ribosomal proteins and elongation factors.[35] UVA-mediated phosphorylation of mTOR and p70S6K may be responsible for an accelerated rate of protein synthesis and activation of tumor cell proliferation. Pomegranate extract has been shown to exhibit an inhibitory effect on UVA-mediated phosphorylation of mTOR with concomitant down-regulation of p70S6K.[26] Rapamycin and its derivatives have been used to block mTOR functions and yield antiproliferative activity in a variety of malignancies. Inhibition of mTOR with resultant G1-S cell-cycle delay and eventual apoptosis along with decreased phosphorylation of p70S6K may be beneficial in the targeting of tumor cells. The mTOR/p70S6K signaling cascade has been recognized to regulate the apoptosis/repair balance initiated by UVB-induced DNA damage and is responsible for the enhanced synthesis of metalloproteinases contributing to connective tissue degradation in photoaging and tumor progression.[36] Pomegranate extract modulation of this pathway in epidermal keratinocytes may have important implications against UVA-mediated damages.

Studies in the mouse model have shown that acute UV exposure results in an apoptotic response through induction of the pro-apoptotic protein Bax with a concomitant decrease in Bcl-2 anti-apoptotic proteins.[37] Bax (-/-) mice showed a significant increase in tumor incidence when exposed to UV-induced stress compared to control mice in two-step chemical carcinogenesis studies.[38] Bcl-X_L, expressed abundantly in the epidermis, prevents cell death induced by UV radiation and other mutagens[39] and its overexpression in transgenic mice dramatically increases the malignant conversion rate of benign tumors.[40] The processes of apoptosis and proliferation in response to UV radiation are closely linked and any dysregulation may lead to the development of skin cancer.[41] A direct link to the apoptosis-regulating proteins has been established through AKT phosphorylation of Bad.[42] Inactivation of AKT prevents it from phosphorylating Bad at Ser[136]. As a result, Bad becomes bound to Bcl-2, increasing its pro-apoptotic activity. Treatment of human epidermal keratinocytes with pomegranate extract prior to UVA exposure resulted in increased expression of the pro-apoptotic proteins Bax and Bad with a concomitant decrease in the anti-apoptotic protein Bcl-X_L, shifting the balance in favor of apoptosis.[26]

Both UVA and UVB act as tumor initiators as well as tumor promoters, but UVB is considered 1,000 to 10,000 times more carcinogenic per J/m^2 than UVA.[43] UVB-induced phosphorylation of the MAPK pathway was shown to be inhibited by the use of antioxidants, thereby suggesting that MAPKs are important targets affected by ROS.[44] Recently, we demonstrated that pretreatment of human epidermal keratinocytes with pomegranate extract resulted in a dose- and time-dependent inhibition of UVB-mediated activation and phosphorylation of MAPK and NFκB pathways.[27] Exposure of NHEK to a physiologically relevant dose of 40 mJ/cm^2 UVB resulted in significant phosphorylation of MAPK proteins. ERK1/2 phosphorylation was seen as early as 15 minutes and was sustained up to 2 hours post-UVB. Pomegranate extract treatment prior to UVB irradiation was found to inhibit UVB-mediated phosphorylation of ERK1/2 in a time-dependent manner. Also, an increased expression of the phosphorylated form of JNK1/2 and p38, evident at 30 minutes post-UVB exposure and persistent until 6 hours, was seen to be markedly reduced in the extract-treated keratinocytes. The distinct level of inhibition of phosphorylation of ERK1/2, JNK1/2, and p38 by PFE at different time points suggests independent mechanisms of regulation of each of these pathways in epidermal keratinocytes.

NFκB activity affects cell survival and determines the sensitivity of cells to cytotoxic agents as well as ionizing radiation. We showed that PFE effectively inhibited UVB-mediated phosphorylation of NFκB/p65 at Ser[536] in the cytoplasm and consequently its nuclear translocation, in a dose- and time-dependent manner.[27] Also a significant decrease in the UVB-mediated NFκB DNA-binding activity was observed in PFE-pretreated keratinocytes. Activation of IKK either directly or indirectly is central to NFκB activation. Pomegranate extract application resulted in significant inhibition of the UVB-mediated increase in phosphorylation of IκBα and IKKα proteins in epidermal keratinocytes. The study shows that the effect of pomegranate extract on NFκB/p65 is mediated through inhibition of phosphorylation and subsequent proteolysis of IκBα. Also, modulation of IKKα by PFE is important because IKK complex is believed to be an important site for integrating signals that regulate the NFκB pathway. Inhibition of UV-mediated proliferative pathways is further evidence of the photochemopreventive role of the extract and its potential to attenuate UV-induced oxidative stress and in turn stress-induced molecular pathways associated with a high risk of carcinogenesis.

Photochemoprevention strategies could be combined with primary preventive strategies for more effective prevention of skin cancer. In this context, a clinical trial showed that topical and oral administration of pomegranate to humans augments the protective effect of sunscreens and affords photoprotection from UVB.[45]

7.6 POMEGRANATE AND PROSTATE CANCER

Prostate cancer is the second-leading cause of cancer-related deaths among men in the U.S. with a similar trend existing in many Western countries.[1] For the year 2005, it has been estimated that 232,090 new cases of prostatic cancer will be diagnosed in men and 30,350 related deaths will occur in the U.S. alone.[1] Although the treatment of localized prostatic cancer remains controversial, there are potentially curative options such as radical prostatectomy or radiotherapy. However, once the disease is metastatic, the outlook for the patient is poor. Chemoprevention is ideally suited for this type of cancer as it is typically diagnosed in men over 50 years of age, and thus even a modest delay in disease progression could significantly impact the quality of life of these patients.[46] In this regard, several naturally occurring antioxidants are being evaluated in cell culture and in animal model systems.[5] Some of these agents like red wine and green tea are showing great promise in prostate cancer patients.[47]

Our recent study showed that PFE exhibited significant antiproliferative and pro-apoptotic activity against highly aggressive human prostate cancer cells (PC3), with a dose-dependent inhibition of cell growth and alterations in the regulatory molecules operative in the G1 phase of the cell cycle.[48] In eukaryotes, passage through the cell cycle is orchestrated by the function of a family of protein kinase complexes.[49] Each complex is composed minimally of a catalytic subunit, the cyclin-dependent kinase or the cdk, and its essential activating partner, the cyclin.[49] Cyclins D and E are involved during $G_1 \rightarrow S$ phase of the cell cycle. In controlled cell growth, association of cyclins D and E with CDKs 2, 4, or 6 leads to phosphorylation of Rb; hyper-phosphorylated Rb leads to its release from E2F. The free E2F then activates c-*myc*, resulting in progression of the cell cycle and cellular proliferation. Any defect in this machinery results in an altered cell cycle regulation with unwanted cellular proliferation culminating in the development of cancer.[49] During the progression of the cell cycle, the CDK-cyclin complexes are inhibited via binding to CDK inhibitory proteins (CDIs) such as WAF1 and KIP1.[50] Prostate cancer PC3 cells treated with PFE showed a significant up-regulation of WAF1/p21 and KIP1/p27 during G_1-phase arrest, independent of p53, with a concomitant down-modulation of the cyclins D1, D2, and E and CDKs 2, 4, and 6, operative in the G_1 phase of the cell cycle.[48] Overexpression of Cyclin D1 and loss of expression of the Rb protein has been reported in several types of cancers; therefore these results signify a critical role of pomegranate extract in the perturbed cell cycle of cancerous cells.

Members of the Bcl-2 family of proteins are critical regulators of the apoptotic pathway. Bcl-2 is an upstream effector molecule in the apoptotic pathway and has been identified as a potent suppressor of apoptosis.[51] It is found at inappropriately high levels in more than half of all human tumors.[51] Bcl-2 has been shown to form a heterodimer complex with the proapoptotic member Bax, thereby neutralizing its proapoptotic effects. Thus, the ratio of Bax/Bcl-2 is considered to be a decisive factor in determining the fate of cells. PFE treatment of PC3 cells was found to result in a decrease in Bcl-2 protein expression with an increase in the protein expression of Bax, altering the ratio of Bax to Bcl-2 in favor of apoptosis.[48] This suggests that up-regulation of Bax and down-modulation of Bcl-2 may be another molecular mechanism through which PFE is capable of inducing apoptosis in prostate cancer cells. This is significant because apoptosis is a physiological process that functions as an essential mechanism of tissue homeostasis and is regarded as the preferred way to eliminate unwanted cells.[5] Moreover, to establish the relevance of these *in vitro* findings to an *in vivo* situation, athymic nude mice were implanted with androgen-responsive CWR22R1 cells, a prostate cancer cell line known to secrete PSA in the bloodstream of the host. Continuous oral feeding of PFE in drinking water, the equivalent of juice intake from two pomegranates by an adult human (approximately 8 oz or 240 ml), significantly slowed the progression of tumor growth in these mice. Importantly, this tumor growth inhibition followed a significant decrease in the serum levels of PSA, a clinical diagnostic serum marker for monitoring the presence and progression of prostate cancer in human patients.[48]

There is evidence that ellagic acid, caffeic acid, luteolin, and punicic acid, all vital components of pomegranate fruit, synergistically inhibit the proliferation and invasion of PC-3 cancer cells across the Matrigel.[52] Pomegranate seed oil as well as polyphenols present in the pericarp and fermented juice potently suppressed proliferation and invasion of several human prostate cancer cells, LNCaP, PC-3, and DU 145, across the Matrigel. In addition, significant inhibition of PC-3 cells was also seen in xenograft-implanted athymic nude mice.[53] Collectively, these studies suggest that pomegranate extract is capable of inhibiting the growth of prostate cancer cells in both *in vitro* and *in vivo* situations. The implications of these findings in cell culture and animal model studies needs to be examined in human prostate cancer patients. This is discussed separately in Chapter 8.

7.7 POMEGRANATE AND LUNG CANCER

Lung cancer is one of the leading causes of cancer-related mortality in the U.S., accounting for 28% of all U.S. cancer deaths, according to the American Cancer Society. Despite recent advances in radiotherapy and chemotherapy, the severe morbidity from lung cancer and the low 5-year survival rates have not improved.[54] The decline is all the more dramatic after the cancer has spread to other organs (16% at regional sites compared to 2.1% at distant sites), making early detection critical. Although several clinical trials are underway, no screening test is yet available for lung cancer that has been proven to improve the survival rates. Cancer chemoprevention is therefore a logical and obvious strategy to help alleviate the incidence and effects of the disease. Several natural compounds, especially plant products and dietary constituents, have been used for its chemoprevention in both *in vitro* and *in vivo* model systems.[55,56] Their mechanisms of action vary widely, with many suppressing cell growth or modulating cell differentiation and some inducing apoptosis.

A relative assessment of the antioxidant capacity of the extract in human lung carcinoma A549 cells showed that PFE possessed higher activity than (-)-epigallo-catechin-3-gallate, the major polyphenol present in green tea.[57] Interestingly, PFE treatment was found to result in a dose-dependent decrease in the viability of A549 cancer cells with only minimal effects on normal human bronchial cells.[57] Also, the effect of PFE on the cell cycle regulatory molecules operative in the G_1 phase was examined. Pretreatment of A549 cells with PFE resulted in marked induction of WAF1/p21 and KIP1/p27 in a dose-dependent manner. Protein expression of cyclins D1, D2, and E and CDKs 2, 4, and 6, regulated by WAF1/p21, also showed a dose-dependent inhibition. The study establishes the involvement of *cki-cyclin-cdk* network in PFE-treated lung cancer cells with resultant inhibition of cell growth and blockade of the cell cycle in a dose-dependent manner. These are important observations because regulation of the cell cycle and apoptosis has become an appreciated target for intervention against cancer.[58] PCNA and Ki67 are cell proliferation markers expressed by actively proliferating cells that undergo rapid degradation as the cell enters the nonproliferative stage. PFE treatment of A549 cells resulted in down-modulation of Ki67 and PCNA protein levels, suggesting that the decrease in growth and viability of lung cancer cells is a result of decreased cellular proliferation. Taken together, these findings suggest that PFE may have chemopreventive as well as cancer chemotherapeutic effects against lung cancer in humans.

A summary of chemopreventive and chemotherapeutic effects of pomegranate and its products is presented in Table 7.1 below. Because of the encouraging data accumulated so far, more detailed studies on the potential of pomegranate-derived products for chemoprevention of lung cancer are warranted.

7.8 FUTURE DIRECTIONS AND CONCLUSIONS

The development of effective and acceptable chemoprevention strategies against human cancers, particularly of the prostate, lung, and skin, has become a medical priority nationally and is also of considerable economic importance. The outcome of these studies could have a direct practical implication and translational relevance to cancer patients if consumption of fruits like pomegranate can retard the process of carcinogenesis and consequently prolong the survival and quality of life of the patient. Our initial research suggests that the pomegranate fruit and its associated antioxidants may possess a strong potential for development as a chemopreventive and possibly as a therapeutic agent against cancers of the skin, lung, and prostate. Our data also indicate that pomegranate extract has a protective role against the adverse effects of UVB- as well as UVA-mediated damages in the culture system, although the mechanisms are not well understood. This provides a framework in which to conduct in-depth molecular studies and to elucidate the components in the fruit that are responsible for its photochemopreventive effects. One might envision the use of such substances in an emollient or patch for chemoprevention or treatment of skin cancer and other skin disorders. Also, *in vivo* studies suggest that pomegranate-derived products are capable of inhibiting conventional as well as novel biomarkers of TPA-induced tumor promotion. In addition, we provided data indicating the potential of pomegranate-derived products for prevention and treatment of lung

TABLE 7.1

Summary of the Reported Chemopreventive and Chemotherapeutic Effects of Pomegranate-Derived Products in Cancers of the Skin, Prostate, and Lung

	Cell Culture	Animal Models	Target/Mechanism(s)	Reference
Skin	NHEK		Inhibits UVA mediated phosporylation of STAT3, AKT, ERK1/2, mTOR, and p70S6K	26
			Upregulates Bax and Bad with concomitant downregulation of Bcl-X_L	
	NHEK		Inhibits UVB-mediated phosporylation of MAPKs and NF-κB pathways	27
		CD-1 mice	Inhibits TPA-mediated increase in skin edema and hyperplasia	8
			Inhibits TPA-induced epidermal ODC activity	8, 23
			Decreases TPA-induced expression of COX-2	
			Inhibits TPA-mediated phosphorylation of MAPKs and NF-κB pathways	
			Decreases tumor incidence and tumor multiplicity	8, 23
		Wistar rats	Accelerates wound healing with increased hydroxyproline content	24
Prostate	PC3		Induces p21 and p27 protein expressions	48
			Downmodulates *cki-cyclin-cdk* network	
			Decreases Bcl-2 protein with an increase in Bax	
	PC3, LnCaP, DU145		Inhibits proliferation and invasion of cancer cells	52, 53
		Athymic nude mice	Inhibits tumor growth and decreases serum PSA levels	48
Lung	A549		Decreases viability of A549 cells	57
			Induces p21 and p27 protein expressions	
			Downmodulates *cki-cyclin-cdk* network	
			Downmodulates proliferation markers such as PCNA and Ki67	

and prostate cancer. Thus it may possess anticancer activity in other *in vivo* situations that must be carefully chosen to evaluate the efficacy of the fruit extract and increase our understanding of its antitumor-promoting effect in diverse model systems. Furthermore, several issues must be addressed, such as determining the optimal dosing and toxicity (if any) of the agent, deciding whether to use a single or combinatorial approach, determining the optimal period and route of administration, and assessing the bioavailability of the active principle. Finally, preclinical trials should be designed to carefully evaluate the tissue-specific effect of pomegranate fruit products and to assess the risk–benefit ratio using complementary strategies. If convincing preclinical data emerges, then clinical trials in high-risk patients must be undertaken.

REFERENCES

1. Jemal, A. et al., Cancer statistics, *CA Cancer J. Clin.,* 55, 10, 2005.
2. Surh, Y.J., Cancer chemoprevention with dietary phytochemicals, *Nat. Rev. Cancer,* 10, 768, 2003.
3. Longtin, R., The pomegranate: nature's power fruit?, *J. Natl. Cancer Inst.,* 95, 346, 2003.
4. Lee, K.W. et al., Vitamin C and cancer chemoprevention: reappraisal, *Am. J. Clin. Nutr.,* 78, 1074, 2003.
5. Mukhtar, H. and Ahmad, N., Cancer chemoprevention: future holds in multiple agents, *Toxicol. Appl. Pharmacol.,* 158, 207, 1999.
6. Katiyar, S.K., Ahmad, N., and Mukhtar, H., Green tea and skin, *Arch. Dermatol.,* 136, 989, 2000.
7. Ahmad, N. et al., Green tea constituent epigallocatechin-3-gallate and induction of apoptosis and cell cycle arrest in human carcinoma cells, *J. Natl. Cancer Inst.,* 89, 1881, 1997.
8. Afaq, F. et al., Anthocyanin- and hydrolyzable tannin-rich pomegranate fruit extract modulates MAPK and NF-kappaB pathways and inhibits skin tumorigenesis in CD-1 mice, *Int. J. Cancer.,* 113, 423, 2005.
9. Halvorsen, B.L. et al., A systematic screening of total antioxidants in dietary plants, *J. Nutr.,* 132, 461, 2002.
10. Alekperov, U.K., Plant antimutagens and their mixtures in inhibition of genotoxic effects of xenobiotics and aging processes, *Eur. J. Cancer Prev.,* 11 Suppl 2, S8, 2002.
11. Singh, R.P., Chidambara Murthy, K.N., and Jayaprakasha, G.K., Studies on the antioxidant activity of pomegranate *(Punica granatum)* peel and seed extracts using *in vitro* models, *J. Agric. Food Chem.,* 50, 81, 2002.
12. Xu, J. et al., Intervention of antioxidant system function of aged rats by giving fruit juices with different antioxidant capacities, *Zhonghua Yu Fang Yi Xue Za Zhi,* 39, 80, 2005.
13. Kawaii, S. and Lansky, E.P., Differentiation-promoting activity of pomegranate *(Punica granatum)* fruit extracts in HL-60 human promyelocytic leukemia cells, *J. Med. Food,* 7, 13, 2004.
14. Gil, M.I. et al., Antioxidant activity of pomegranate juice and its relationship with phenolic composition and processing, *J. Agric. Food Chem.,* 10, 4581, 2000.
15. Verma, A.K. et al., Correlation of the inhibition by retinoids of tumor promoter-induced mouse epidermal ornithine decarboxylase activity and of skin tumor promotion, *Cancer Res.,* 39, 419, 1979.
16. Janne, J. et al., Genetic approaches to the cellular functions of polyamines in mammals, *Eur. J. Biochem.,* 271, 877, 2004.
17. Fletcher, B.S. et al., Structure of the mitogen-inducible TIS10 gene and demonstration that the TIS10-encoded protein is a functional prostaglandin G/H synthase, *J. Biol. Chem.,* 267, 4338, 1992.
18. Meade, E.A., Smith, W.L., and DeWitt, D.L., Differential inhibition of prostaglandin endoperoxide synthase (cyclooxygenase) isozymes by aspirin and other non-steroidal anti-inflammatory drugs, *J. Biol. Chem.,* 268, 6610, 1993.

19. Prescott, S.M. and Fitzpatrick F.A., Cyclooxygenase-2 and carcinogenesis, *Biochim. Biophys. Acta.*, 1470, M69, 2000.

20. Ding, M. et al., Inhibition of AP-1 and neoplastic transformation by fresh apple peel extract, *J. Biol. Chem.*, 279, 10670, 2004.

21. Carter, A.B. et al., The p38 mitogen-activated protein kinase is required for NF-kappaB-dependent gene expression. The role of TATA-binding protein (TBP), *J. Biol. Chem.*, 274, 30858, 1999.

22. Garg, A. and Aggarwal, B.B., Nuclear transcription factor-kappaB as a target for cancer drug development, *Leukemia,* 16, 1053, 2002.

23. Hora, J.J. et al., Chemopreventive effects of pomegranate seed oil on skin tumor development in CD1 mice, *J. Med. Food,* 6, 157, 2003.

24. Murthy, K.N. et al., Study on wound healing activity of *Punica granatum* peel, *J. Med. Food,* 7, 256, 2004.

25. Afaq, F. et al., Botanical antioxidants for chemoprevention of photocarcinogenesis, *Front. Biosci.,* 7, d784, 2002.

26. Syed, D.N. et al., Photochemopreventive effect of pomegranate fruit extract on UVA-mediated activation of cellular pathways in normal human epidermal keratinocytes, *Photochem. Photobiol.,* (In press), 2005.

27. Afaq, F. et al., Pomegranate fruit extract modulates UV-B-mediated phosphorylation of mitogen-activated protein kinases and activation of nuclear factor kappa B in normal human epidermal keratinocytes, *Photochem. Photobiol.,* 81, 38, 2005.

28. Bowden, G.T., Prevention of non-melanoma skin cancer by targeting ultraviolet-B-light signaling, *Nat. Rev. Cancer,* 4, 23, 2004.

29. Afaq, F., Adhami, V.M., and Mukhtar, H., Photochemoprevention of ultraviolet B signaling and photocarcinogenesis, *Mutat. Res.,* 571, 153, 2005.

30. Levy, D.E. and Darnell, J.E., Stats: transcriptional control and biological impact. *Nat. Rev. Mol. Cell Biol.,* 3, 651, 2002.

31. Chan, K.S. et al., Disruption of Stat3 reveals a critical role in both the initiation and the promotion stages of epithelial carcinogenesis, *J. Clin. Invest.,* 114, 720, 2004.

32. Zykova, T.A. et al., The signal transduction networks required for phosphorylation of STAT1 at Ser727 in mouse epidermal JB6 cells in the UVB response and inhibitory mechanisms of tea polyphenols, *Carcinogenesis,* 26, 331, 2005.

33. Hildesheim, J. and Fornace, A.J., The dark side of light: the damaging effects of UV rays and the protective efforts of MAP kinase signaling in the epidermis, *DNA Repair,* 3, 567, 2004.

34. He, Y.Y., Huang, J.L., and Chignell, C.F., Delayed and sustained activation of extra-cellular signal-regulated kinase in human keratinocytes by UVA: implications in carcinogenesis, *J. Biol. Chem.,* 279, 53867, 2004.

35. Stewart, M.J. and Thomas, G., Mitogenesis and protein synthesis: a role for ribosomal protein S6 phosphorylation? *Bioessays,* 16, 809, 1994.

36. Brenneisen, P. et al., Activation of p70 ribosomal protein S6 kinase is an essential step in the DNA damage-dependent signaling pathway responsible for the ultraviolet B-mediated increase in interstitial collagenase (MMP-1) and stromelysin-1 (MMP-3) protein levels in human dermal fibroblasts, *J. Biol. Chem.,* 275, 4336, 2000.

37. Melnikova, V.O. and Ananthaswamy, H.N., Cellular and molecular events leading to the development of skin cancer, *Mutat. Res.,* 571, 91, 2005.

38. Cho, S.H. et al., Bax gene disruption alters the epidermal response to ultraviolet irradiation and *in vivo* induced skin carcinogenesis, *Int. J. Mol. Med.,* 7, 235, 2001.

39. Krajewski, S. et al., Immunohistochemical analysis of *in vivo* patterns of Bcl-X$_L$ expression, *Cancer Res.,* 54, 5501, 1994.

40. Pena, J.C., Rudin, C.M., and Thompson, C.B., A Bcl-xL transgene promotes malignant conversion of chemically initiated skin papillomas, *Cancer Res.,* 58, 2111, 1998.
41. Ouhtit, A. et al., Temporal events in skin injury and the early adaptive responses in ultraviolet-irradiated mouse skin, *Am. J. Pathol.,* 156, 201, 2000.
42. Chun, K.S. et al., Effects of yakuchinone A and yakuchinone B on the phorbol ester-induced expression of COX-2 and iNOS and activation of NF-kappaB in mouse skin, *J. Environ. Pathol. Toxicol. Oncol.,* 21, 131, 2002.
43. Bachelor, M.A. and Bowden, G.T., UVA-mediated activation of signaling pathways involved in skin tumor promotion and progression, *Semin. Cancer Biol.,* 14, 131, 2004.
44. Katiyar, S.K. et al., Inhibition of UVB-induced oxidative stress-mediated phosphorylation of mitogen-activated protein kinase signaling pathways in cultured human epidermal keratinocytes by green tea polyphenol (-)-epigallocatechin-3-gallate, *Toxicol. Appl. Pharmacol.,* 176, 110, 2001.
45. Murad, H. and Shellow, V.R.W., Pomegranate extract both orally ingested and topically applied to augment the SPF of sunscreens, *Cosmet. Dermatol.,* 14, 43, 2001.
46. Saleem, M. et al., Tea beverage in chemoprevention of prostate cancer, *Nutr. Cancer,* 47, 13, 2003.
47. Greenwald, P., Lifestyle and medical approaches to cancer prevention, *Recent Results Cancer Res.,* 166, 1, 2005.
48. Malik, A. et al., Pomegranate fruit juice for chemoprevention and chemotherapy of prostate cancer, *Proc. Natl. Acad. Sci.,* (In Press), 2005.
49. Sanchez, I. and Dynlacht, B.D., New insights into cyclins, CDKs, and cell cycle control, *Semin. Cell Dev. Biol.,* 16, 311, 2005.
50. Coqueret, O., New roles for p21 and p27 cell-cycle inhibitors: a function for each cell compartment?, *Trends Cell Biol.,* 13, 65, 2003.
51. Gandour-Edwards, R. et al., Abnormalities of apoptotic and cell cycle regulatory proteins in distinct histopathologic components of benign prostatic hyperplasia, *Prostate Cancer Prostatic Dis.,* 7, 321, 2004.
52. Lansky, E.P. et al., Pomegranate *(Punica granatum)* pure chemicals show possible synergistic inhibition of human PC-3 prostate cancer cell invasion across Matrigel, *Invest. New Drugs,* 23, 121, 2005.
53. Albrecht, M. et al., Pomegranate extracts potently suppress proliferation, xenograft growth, and invasion of human prostate cancer cells, *J. Med. Food,* 7, 274, 2004.
54. Khuri, F.R., Herbst, R.S., and Fossella, F.V., Emerging therapies in non-small-cell lung cancer, *Ann. Oncol.,* 12, 739, 2001.
55. Pezzuto, J.M., Plant-derived anticancer agents, *Biochem. Pharmacol.,* 53, 121, 1997.
56. Kelloff, G.J. et al., Progress in cancer chemoprevention: development of diet derived chemopreventive agents, *J. Nutr.,* 130, 467S, 2000.
57. Hadi, N. et al., Antiproliferative effects of anthocyanins, hydrolyzable and oligomeric tannins rich pomegranate fruit extract on human lung carcinoma cells A549, *Proc. Am. Assoc. Cancer Res.,* 46, 579, 2005.
58. Ahmad, N., Gupta, S., and Mukhtar, H., Involvement of retinoblastoma (Rb) and E2F transcription factors during photodynamic therapy of human epidermoid carcinoma cells A431, *Oncogene,* 18, 1891, 1999.

8 Pomegranate and Prostate Cancer Chemoprevention

John T. Leppert and Allan J. Pantuck

CONTENTS

8.1 INTRODUCTION

Adenocarcinoma of the prostate is currently the most common malignancy in men in the U.S., comprising 29% of all cancers. This year an estimated 232,090 men will be newly diagnosed with prostate cancer.[1] The incidence of prostate cancer demonstrates a striking rise with increasing age,[2] and the number of men older than 65 is expected to double by 2020. There has been a trend toward improved survival in prostate cancer patients, and prostate cancer 5-year survival rates have increased from 67% for the period 1974 to 1976 to 92% for the period 1989 to 1995.[3] This improvement is likely due to multiple factors, including better screening for prostate cancer resulting in stage migration to earlier diagnosis as well as more effective therapy. However, prostate cancer remains the second most common cause of cancer death in men in the U.S., accounting for 11% of all cancer deaths. This year an estimated 30,350 men in America,[1] and 221,002 worldwide,[4] will die of prostate cancer. These data underscore the medical and socioeconomic importance of prostate cancer and the development of effective treatment strategies.

8.2 TREATMENT OF LOCALIZED PROSTATE CANCER

Primary management of prostate cancer for the majority of patients consists either of radical surgery or radiation therapy. Although this is adequate for permanent disease control in many patients, a significant number of patients relapse and ultimately develop metastatic disease. Radical prostatectomy is currently the most commonly utilized therapy for curative intent.[5] However, approximately one-third of prostate cancer patients with clinically confined cancer who are treated with radical prostatectomy will develop a biochemical recurrence of disease.[6,7] Prostate-specific antigen (PSA) is a single-chain glycoprotein produced by the epithelial cells of the prostate that has been used for early detecting and staging of tumors and for monitoring patients with prostate cancer of different stages who receive a variety of treatments.[8] Pound et al.[9] reviewed 1997 patients who underwent radical prostatectomy for clinically localized prostate cancer and determined that 15% of patients had a biochemical recurrence, defined as a positive PSA test following treatment. In this study, 34% of patients with biochemical recurrence developed distant metastases within 15 years of total follow-up, median time to development of metastases was 8 years from the time of initial PSA elevation, and median time to death from the development of metastases was 5 years.[9]

8.3 TREATMENT OF METASTATIC PROSTATE CANCER

Historically, management of metastatic prostate cancer has relied primarily on hormonal therapy. This has taken the form of orchiectomy, estrogens, leutinizing hormone-releasing hormone (LHRH) agonists, or antiandrogens. These hormonal manipulations have provided significant palliation in men with metastatic disease but are not curative. Due to the widespread use of serum PSA to monitor for disease recurrence following primary treatment for prostate cancer, many patients are being diagnosed with rising PSA and no clinical or radiographic evidence of recurrence. In this clinical setting, therapeutic options are limited, tumor burden is relatively low and where patients are suitable for clinical trials. Currently there are limited treatment options for these patients, none of which has been proven to improve survival. Furthermore, early initiation of hormonal ablation is associated with significant morbidity and impact on quality of life, including fatigue, hot flashes, loss of libido, decreased muscle mass, and osteoporosis with long-term use. Currently these patients generally are managed through observation until their clinical course dictates hormonal intervention, and the therapeutic option of hormonal therapy remains available to these patients if they develop overt metastatic disease. Strategies to delay clinical prostate cancer progression and prolong the interval from treatment failure to hormonal ablation would be of paramount importance.

8.4 NUTRITIONAL CANCER CHEMOPREVENTION

The field of chemoprevention lies at an interesting intersection between disease management and health promotion. It is believed that the process of carcinogenesis results from the interaction of environmental exposures, including diet, and genetic susceptibility. The molecular pathology of cancer begins not with the appearance of clinically apparent lesions, but rather with altered biochemical and genetic processes. The evidence reviewed below suggests that the contents of one's diet may impact the development and progression of cancer. Many authorities, however, question the validity of current recommendations for nutritional chemoprevention against cancer. The reason for this skepticism revolves around the wide variations reported in epidemiologic studies that are the nature of observational studies. Observational dietary studies have been limited in their conclusions because the protection afforded by the consumption of a particular nutrient may be multifactorial, with different components of the food exerting potential chemopreventive effects. Furthermore, measuring levels of nutrients in the food intake of populations is confounded by factors that might affect these levels and also the incidence of cancer. In addition, chemoprevention studies using dietary strategies may be expected to have mild effects, and large studies may be required to confirm statistical significance. Therefore, prospective trials with a sufficient sample size, ample follow-up, an extended duration of treatment, and validated biomarkers of risk, effect, and prognosis are needed to clarify the association between micronutrients and cancer protection.

8.5 CHEMOPREVENTION OF PROSTATE CANCER

Age, race, and family history are the only established risk factors for prostate cancer. However, research efforts are investigating genetic and environmental factors involved in the development of prostate cancer. The incidence of prostate cancer varies tremendously by geographic location.[4] Evidence continues to mount that environmental factors, particularly diet, play a pivotal role in the development of prostate cancer. It has been known for some time that patients migrating from low-risk to high-risk locations acquire the cancer risk of their new location.[10] Identifying the specific micronutrients that lead to, or protect against, the development of prostate cancer has the potential to significantly alter the approach to prostate cancer treatment and prevention.

The last 10 years have witnessed the ability of chemoprevention to reduce mortality associated with common epithelial cancers.[11] A combination of epidemiologic and basic science evidence strongly suggests that diet and plant-derived phytochemicals may play an important role in prostate cancer prevention or treatment. Japanese and Chinese native to their countries who consume a low-fat and high-fiber diet with high consumption of phytochemicals that include soy products and green tea have the lowest incidence of prostate cancer.[12] Epidemiological studies suggest that a reduced risk of cancer is associated with the consumption of a phytochemical-rich diet that includes fruits and vegetables.[13] Fresh and processed fruits and food products contain high levels of a diverse range of phytochemicals of which polyphenols, including hydrolysable tannins (ellagitannins [ETs] and gallotannins) and condensed tannins (proanthocyanidins), and anthocyanins and other flavonoids make up a large proportion.[14] A number of phytochemicals, including resveratrol from grapes and red wine, sulforaphane from broccoli and other cruciferous vegetables, organosulfides from garlic and other allium species, limonene and perillyl alcohol from the lipid fraction of citrus peels, isoflavones such as genistein and daidzein and enterodiols from soy and flax proteins, catechins from green tea, and lycopene from tomatoes have been proposed as potential chemoprevention agents based on animal and laboratory evidence of antitumor effects. Suggested mechanisms of anticancer effects of polyphenols include, in addition to their role as potent antioxidants, (1) the inhibition of human cancer cell growth by interfering with growth factor receptor signaling and cell cycle progression, (2) promotion of cellular differentiation, (3) induction of hepatic xenobiotic enzyme activities that may provide additional defense mechanisms against oxidant stress and carcinogens, (4) inhibition of cholesterogenesis, (5) modulation of phosphodiesterase and cyclooxygenase pathways, (6) inhibition of protein kinases involved in cell signaling, and (7) inhibition of inflammation.[15–17]

8.6 POMEGRANATE AND PROSTATE CANCER CHEMOPREVENTION

The pomegranate (*Punica granatum* L.) fruit has been used for centuries in ancient cultures for its medicinal purposes.[18] Pomegranate fruits are widely consumed fresh and in beverage forms as juice and wines. Commercial pomegranate juice shows potent antioxidant and antiatherosclerotic properties attributed to its high content of polyphenols including ellagic acid (EA) in its free and bound forms (as ETs and EA-glycosides [EAGs]), gallotannins, and anthocyanins (cyanidin, delphinidin, and pelargonidin glycosides) and other flavonoids (quercetin, kaempferol, and luteolin glycosides).[19] The most abundant of these polyphenols is punicalagin, an ET implicated as the bioactive constituent responsible for >50% of the juice's potent antioxidant activity. Punicalagin is abundant in the fruit husk and during processing is extracted into pomegranate juice in significant quantities, reaching levels of >2 g/L juice.[19] As a result, pomegranate juice displays higher antioxidant activity than green tea and red wine.[19]

The micronutrients in pomegranate juice have demonstrated anticancer activity in studies of breast cancer,[20] leukemia,[21] and prostate cancer cells *in vitro*.[22–26] Albrecht et al.[22] studied the effect of different fractions of pomegranate juice against the prostate cell lines LNCaP, PC-3, and DU 145. Both fermented juice and the extracted polyphenol fractions were potent inhibitors of proliferation of these prostate cancer cell lines. The LNCaP cell line, with functional androgen receptors, was particularly sensitive to doses of pomegranate extracts. Further examination revealed that after 72 hours of treatment with the pomegranate extracts, the prostate cancer cells demonstrated altered cell-cycle characteristics and began showing between a 5- and 9-fold increase in apoptosis. In addition, pomegranate extracts inhibited the invasion of the PC-3 cell line through a Matrigel system. This data suggests that pomegranate extract interferes with prostate cancer cell growth and metastatic potential, and ultimately led to increased prostate cancer cell death.

Seeram et al.[26] investigated particular phytochemical extracts found in pomegranate juice against oral, colon, and prostate cancer cell lines. Punicalagin, EA, and total pomegranate tannin showed a 40 to 100% reduction in the growth of

RWPE-1 and 22Rv1 prostate cancer cells. The authors also found that commercial pomegranate juice, with several active ingredients, demonstrated more antitumor activity than the specific fractions alone. Lanksy et al.[24] also examined the ability of discrete fractions of pomegranate juice to suppress prostate cancer cell growth. The authors identified an antiproliferative effect of EA, caffeic acid, luteolin, and punicic acid on DU 145 and PC-3 cell lines. Again, the authors identified a synergistic interaction among the individual fractions, with combined fractions demonstrating increased activity.[23] The synergistic effects were also demonstrated as these fractions inhibited the invasion of PC-3 cells through a Matrigel model. A combination of these discrete fractions decreased cell invasion by nearly 100%, while each individual fraction resulted in a 60 to 70% reduction. This suggests that phytochemicals commonly isolated from pomegranate peel, as well as the seeds, may be active against prostate cancer growth.

Recently, Malik et al.[27] published their studies of pomegranate fruit extract (PFE) as a chemopreventive agent for prostate cancer. The authors' report confirmed the ability of PFE to inhibit the cell growth of PC-3 cells. In addition, PFE resulted in increased prostate cancer cell apoptosis in a dose-dependant manner. This report further characterized the molecular signals and pathways activated by exposure to PFE. PFE exposure resulted in an induction of the WAF1/p21 pathways, known to regulate cell division. In addition, this study illustrated altered expression of the Bax and Bcl-2 proteins leading to the increase in apoptosis. This study also provided the first evidence of the activity of PFE on prostate cancer cells in an animal model. PFE postponed the development of prostate cancer in mice injected with CWR22Rv1 cells. Mice given regular water developed tumors approximately 8 days after injection. Mice treated with PFE did not demonstrate evidence of prostate cancer until 11 to 14 days after injection. In addition, the PSA levels were reduced 70 to 85% in mice with diet supplemented with PFE.

These scientific and preclinical studies strongly suggest that particular pomegranate phytochemicals may have activity against prostate cancer growth. We recently reported the first clinical trial of pomegranate juice in human patients with prostate cancer.[28] A phase II clinical trial was performed at UCLA for patients with biochemical recurrence of prostate cancer following surgery or radiation. Patients were treated with 8 ounces of pomegranate juice by mouth daily (Wonderful variety, equivalent to 1.5 mmol of total polyphenols per day). There were no serious adverse events reported and the treatment was well tolerated, and no patients developed metatsatic disease on study. Patients were followed in 3-month intervals, with studies to evaluate serum PSA levels, hormone levels (testosterone, estradiol, sex hormone-binding globulin [SHBG], dehydroepiandrosterone [DHEA], insulin growth factor [IGF], and androstenedione), markers for treatment compliance (serum and urinary polyphenol/ellagic acid levels), markers of serum antioxidant effect (serum nitric oxide levels), and *in vitro* assays that measure the effect of patients' serum on the LNCaP growth and apoptosis.[29] Thirty-five percent of patients achieved a decrease in PSA during treatment. The average PSA doubling time (PSADT), a surrogate marker of cancer progression, increased 15 months (median 11.5 months) to 37 months (median 19.9 months) after pomegranate juice therapy. The 22-month prolongation in PSADT reached statistical significance (p = 0.0001). Among the patients, 83%

achieved an improvement in PSADT (i.e., either prestudy PSADT>0 and poststudy PSADT<0, or post-PSADT>pre-PSADT) after intervention ($p < 0.0001$).

The effect of pomegranate juice treatment on prostate cancer cell growth was also assessed. LNCaP prostate cancer cells were grown in the serum of patients before initiating pomegranate supplementation as well as in serum from patients after 9 months of treatment. These studies corroborate the findings in the preclinical pomegranate studies. Compared to baseline, at 9 months there was a 12% reduction in the growth of LNCaP (p = 0.0048) with 84% of patients demonstrating a month 9 value that was less than their respective baseline. This change in cellular proliferation was associated with a corresponding 17.5% increase in apoptosis at 9 months compared to baseline (p = 0.0004), with 75% of patients tested showing an increase in apoptosis. In addition, nitric oxide metabolites were measured to evaluate the level of antioxidant activity in patients treated with pomegranate juice. Compared to baseline, there was a 23% increase in serum nitric oxide metabolites measured in patients' serum at 9 months ($p = 0.0085$) with two-thirds of patients assayed having an increase compared to baseline. Interestingly, there was no significant difference in pre- and posttreatment patient hormone levels.

8.7 DISCUSSION

Pomegranate juice contains promising phytochemicals for the chemoprevention and chemotherapy of prostate cancer. It is interesting to speculate on potential mechanisms of action of pomegranate polyphenols on the growth dynamics of *in vivo* prostate cancer. Chief among these hypotheses is the role of inflammation in cancer, and the antioxidant and anti-inflammatory effects of pomegranate polyphenols. Chronic inflammation has been linked to the incidence of many cancers, including that of the prostate.[30] An increased risk of cancer is associated with inflammatory mechanisms, as approximately 15% of all cancers can be related to chronic inflammation.[31] Epidemiologic studies have found an increased risk for prostate cancer in men who have a prior history of sexually transmitted disease or prostatitis.[32] Inflammation in the microenvironment of the prostate cancer cell may stimulate the multistep process of carcinogenesis by upregulating the production of pro-inflammatory cytokines and their signaling pathways. In fact, proliferative inflammatory atrophy has recently been proposed as a precursor to prostatic intraepithelial neoplasia (PIN) and prostate cancer.[33]

Inflammation can result in persistent oxidative stress in cancer cells and the reactive oxygen species (ROS) may lend cancer cells a survival advantage.[34–36] Mild levels of oxidative stress stimulate cancer cell proliferation[35] and increase mutation rates through DNA damage or epigenetic changes.[37] Nelson et al.[33] have demonstrated the loss of GSTP-1 as an early event in prostate tumors that sets the stage for stimulation of growth by oxygen radicals. Oxidative stress has also been shown to increase cancer cell proliferation by increasing the sensitivity of growth factor receptors and by altering transcription factor activity. Potent antioxidant and prostaglandin-inhibitory activities have been previously demonstrated for polyphenols extracted from pomegranate juice and seed oil.[38] In skin cancer models, pomegranate seed oil significant decreased tumor incidence and multiplicity in CD_1 mice, and was associated with an inhibition of TPA-induced ornithine decarboxylase activity,[39] as well as modulation of MAPK and NF-κB pathways.[40] The suppression of NF-κB is notable, and it has been noted that pomegranate can suppress NF-κB activation in other models such as vascular endothelial cells.[41] Activation of the NF-κB transcription factor results in up-regulation of antiapoptotic genes, and NF-κB is thought to be a key factor in the control of cell proliferation, inhibition of apoptosis, and oncogenesis in many cancers including prostate cancer.[42]

It has been previously shown that intervention with antioxidants can increase eNOS expression in both cultured endothelial cells and in hypercholesterolemic mice.[43] More recently, it has been reported that polyphenol flavonoids contained in pomegranate juice are capable of eliciting similar effects, enhancing the biological actions of NO by virtue of their capacity to stabilize NO. By protecting against the oxidative destruction of NO by reactive oxygen species and other radicals, pomegranate juice has produced higher and more prolonged cellular NO concentrations and biological actions.[44] The role of NO as a potential chemopreventive agent remains controversial, with both inhibitory and promoting effects on neoplasia being reported secondary to the NO/iNOS system.[45] In studies *in vitro,* nitric-oxide-donating aspirin (NO-ASA), consisting of ASA (aspirin) plus an -ONO(2) moiety linked via a molecular spacer, has been shown to inhibit the growth of colon, pancreatic, lung, skin, leukemia, breast, and prostate cancer cells,[46] owing to both inhibition of cellular proliferation by blocking the G(1) to S cell-cycle transition and to induction of apoptosis, at least partially through inhibition of NF-κB. In the clinical trial of pomegranate juice in patients with prostate cancer,[28] serum NO levels increased as the rate of rise of PSA slowed down, providing indirect support for NO as an antiproliferative in prostate cancer and as one potential mechanism for the biological effects observed.

The long natural history of localized prostate cancer makes it a particularly difficult disease in which to study the impact of therapy. Cancer cure and prolongation of life are the ultimate goals of cancer therapies. However, men with PSA recurrence after surgery often live more than a decade even if they are not cured by their therapy, and it may take over 15 years from a study's inception to demonstrate a survival advantage in this patient group.[9]

While cure for men with clinically localized prostate cancer can frequently be accomplished definitively through surgery, radiation, or chemotherapy, it can also be accomplished in theory by slowing down the progression of the cancer to the point where the patient dies of competing, nonprostate cancer causes.

8.8 CONCLUSIONS

The consumption of pomegranate juice is a promising chemopreventive strategy for patients with prostate cancer. Early reports also suggest that pomegranate juice, acting through antioxidant and anti-inflammatory pathways, may also represent a novel therapeutic agent in the treatment of patients with prostate cancer.

REFERENCES

1. Jemal, A. et al., Cancer statistics, 2005. *CA Cancer J Clin,* 55, 10, 2005.
2. Stanford, J.L. et al., *Prostate cancer trends 1973–1995.* NIH Pub. No. 99-4543: 1999.
3. Greenlee, R.T. et al., Cancer statistics, 2000. *CA Cancer J Clin,* 50, 7, 2000.
4. Parkin, D.M. et al., Global cancer statistics, 2002. *CA Cancer J Clin,* 55, 74, 2005.
5. Petrovich, Z. et al., Adenocarcinoma of the prostate: Innovations in management. *Am J Clin Oncol,* 20, 111, 1997.
6. Trapasso, J.G. et al., The incidence and significance of detectable levels of serum prostate specific antigen after radical prostatectomy. *J Urol,* 152, 1821, 1994.
7. Zincke, H. et al., Long-term (15 years) results after radical prostatectomy for clinically localized (stage t2c or lower) prostate cancer. *J Urol,* 152, 1850, 1994.
8. Catalona, W.J. et al., Measurement of prostate-specific antigen in serum as a screening test for prostate cancer. *N Engl J Med,* 324, 1156, 1991.
9. Pound, C.R. et al., Natural history of progression after PSA elevation following radical prostatectomy. *JAMA,* 281, 1591, 1999.
10. Haenszel, W. and Kurihara, M., Studies of Japanese migrants. I. Mortality from cancer and other diseases among Japanese in the United States. *J Natl Cancer Inst,* 40, 43, 1968.
11. Hong, W.K. and Sporn, M.B., Recent advances in chemoprevention of cancer. *Science,* 278, 1073, 1997.
12. Yip, I., Heber, D., and Aronson, W., Nutrition and prostate cancer. *Urol Clin North Am,* 26, 403, 1999.
13. Block, G., Patterson, B., and Subar, A., Fruit, vegetables, and cancer prevention: A review of the epidemiological evidence. *Nutr Cancer,* 18, 1, 1992.
14. Knekt, P. et al., Flavonoid intake and risk of chronic diseases. *Am J Clin Nutr,* 76, 560, 2002.
15. Bravo, L., Polyphenols: Chemistry, dietary sources, metabolism, and nutritional significance. *Nutr Rev,* 56, 317, 1998.
16. Middleton, E., Jr., Kandaswami, C., and Theoharides, T.C., The effects of plant flavonoids on mammalian cells: Implications for inflammation, heart disease, and cancer. *Pharmacol Rev,* 52, 673, 2000.
17. Yang, C.S. et al., Inhibition of carcinogenesis by dietary polyphenolic compounds. *Annu Rev Nutr,* 21, 381, 2001.
18. Longtin, R., The pomegranate: Nature's power fruit? *J Natl Cancer Inst,* 95, 346, 2003.
19. Gil, M.I. et al., Antioxidant activity of pomegranate juice and its relationship with phenolic composition and processing. *J Agric Food Chem,* 48, 4581, 2000.
20. Kim, N.D. et al., Chemopreventive and adjuvant therapeutic potential of pomegranate *(punica granatum)* for human breast cancer. *Breast Cancer Res Treat,* 71, 203, 2002.
21. Kawaii, S. and Lansky, E.P., Differentiation-promoting activity of pomegranate *(punica granatum)* fruit extracts in hl-60 human promyelocytic leukemia cells. *J Med Food,* 7, 13, 2004.
22. Albrecht, M. et al., Pomegranate extracts potently suppress proliferation, xenograft growth, and invasion of human prostate cancer cells. *J Med Food,* 7, 274, 2004.
23. Lansky, E.P. et al., Pomegranate *(punica granatum)* pure chemicals show possible synergistic inhibition of human pc-3 prostate cancer cell invasion across Matrigel. *Invest New Drugs,* 23, 121, 2005.
24. Lansky, E.P. et al., Possible synergistic prostate cancer suppression by anatomically discrete pomegranate fractions. *Invest New Drugs,* 23, 11, 2005.

25. Mukhtar H. and Syed, D., In *Pomegranate fruit exract for chemprevention and chemotherapy of prostate cancer*, AACR Frontiers in Cancer Prevention Research, Seattle, Washington, 2004.

26. Seeram, N.P. et al., *In vitro* antiproliferative, apoptotic and antioxidant activities of punicalagin, ellagic acid and a total pomegranate tannin extract are enhanced in combination with other polyphenols as found in pomegranate juice. *J Nutr Biochem,* 16, 360, 2005.

27. Malik, A. et al., Pomegranate fruit juice for chemoprevention and chemotherapy of prostate cancer. *Proc Natl Acad Sci USA,* 102, 14813, 2005.

28. Pantuck, A.J. et al., Phase II study of pomegranate juice for men with rising PSA following surgery or radiation for prostate cancer. *Clin. Cancer Res.,* 2006, in press.

29. Tymchuk, C.N. et al., Evidence of an inhibitory effect of diet and exercise on prostate cancer cell growth. *J Urol,* 166, 1185, 2001.

30. Weitzman, S.A. and Gordon, L.I., Inflammation and cancer: Role of phagocyte-generated oxidants in carcinogenesis. *Blood,* 76, 655, 1990.

31. Kuper, H., Adami, H.O., and Trichopoulos, D., Infections as a major preventable cause of human cancer. *J Intern Med,* 248, 171, 2000.

32. Palapattu, G.S. et al., Prostate carcinogenesis and inflammation: Emerging insights. *Carcinogenesis,* 26, 1170, 2005.

33. De Marzo, A.M. et al., Human prostate cancer precursors and pathobiology. *Urology,* 62, 55, 2003.

34. Dreher, D. and Junod, A.F., Role of oxygen free radicals in cancer development. *Eur J Cancer,* 32A, 30, 1996.

35. Kondo, S. et al., Persistent oxidative stress in human colorectal carcinoma, but not in adenoma. *Free Radic Biol Med,* 27, 401, 1999.

36. Toyokuni, S. et al., Persistent oxidative stress in cancer. *FEBS Lett,* 358, 1, 1995.

37. Wainfan, E. and Poirier, L.A., Methyl groups in carcinogenesis: Effects on DNA methylation and gene expression. *Cancer Res,* 52, 2071s, 1992.

38. Schubert, S.Y., Lansky, E.P., and Neeman, I., Antioxidant and eicosanoid enzyme inhibition properties of pomegranate seed oil and fermented juice flavonoids. *J Ethnopharmacol,* 66, 11, 1999.

39. Hora, J.J. et al., Chemopreventive effects of pomegranate seed oil on skin tumor development in cd1 mice. *J Med Food,* 6, 157, 2003.

40. Afaq, F. et al., Anthocyanin- and hydrolyzable tannin-rich pomegranate fruit extract modulates MAPK and NF-kappaB pathways and inhibits skin tumorigenesis in CD-1 mice. *Int J Cancer,* 113, 423, 2005.

41. Schubert, S.Y., Neeman, I., and Resnick, N., A novel mechanism for the inhibition of NF-kappaB activation in vascular endothelial cells by natural antioxidants. *FASEB J,* 16, 1931, 2002.

42. Shukla, S. et al., Constitutive activation of p i3 k-AKT and NF-kappaB during prostate cancer progression in autochthonous transgenic mouse model. *Prostate,* 64, 224, 2005.

43. De Nigris, F. et al., Beneficial effects of antioxidants and l-arginine on oxidation-sensitive gene expression and endothelial no synthase activity at sites of disturbed shear stress. *Proc Natl Acad Sci USA,* 100, 1420, 2003.

44. De Nigris, F. et al., Beneficial effects of pomegranate juice on oxidation-sensitive genes and endothelial nitric oxide synthase activity at sites of perturbed shear stress. *Proc Natl Acad Sci USA,* 102, 4896, 2005.

45. Crowell, J.A. et al., Is inducible nitric oxide synthase a target for chemoprevention? *Mol Cancer Ther,* 2, 815, 2003.
46. Kashfi, K. and Rigas, B., Molecular targets of nitric-oxide-donating aspirin in cancer. *Biochem Soc Trans,* 33, 701, 2005.

9 Assessment of Estrogenicity of Pomegranate in an *In Vitro* Bioassay

Diane M. Harris, Emily Besselink, and Navindra P. Seeram

CONTENTS

9.1 INTRODUCTION

The pomegranate *(Punica granatum)* has been held sacred by many of the world's major religions. A major theme of the pomegranate in these historical traditions is a symbol of life and regeneration. In an explanation of the choice of the pomegranate as the official logo of the British Medical Association's Millennium Festival of Medicine, Langley reviews how the pomegranate is featured on the heraldic crests of several British medical institutions, due to its medicinal properties.[1] History illustrates that for a number of cultures, the red, glistening, juice-encapsulated pomegranates were perceived as a symbol for fertility and fecundity; the shape of the fruit itself is seen as reminiscent of a female breast.[2] Ethnomedical explorations have shown that pomegranate hull and/or root extract has been used orally and intravaginally to prevent both fertility and abortion and to treat various gynecological conditions, predicting the presence of compounds with hormonal activity.[2] In modern knowledge it is known that pomegranate extracts contain many potentially estogenic compounds, thus apparently, as has often been described, traditional knowledge of medicinal properties of plants predicts modern analyses.

Estrogens are a group of steroid hormones that function as the primary female sex hormones. Three compounds, 17β-estradiol (E_2), estriol (E_3), and estrone (E_1), are C_{18} steroids derived from cholesterol, and account for most of the estrogenic activity in humans. While estrogens are present in both men and women, they are found in women in significantly higher quantities. They promote the development of female secondary sexual characteristics, such as breasts, and are also involved in controlling the menstrual cycle; most oral contraceptives contain estrogens. They are produced primarily by developing follicles in the ovaries, the corpus luteum, and the placenta, although some estrogens are also produced in other tissues such as liver, adrenal glands, and adipose tissue. The major estrogen secreted by the ovary is 17β-estradiol, which is the most bioactive estrogen. In women, estrone and estradiol are synthesized from androstenedione and testosterone, respectively, by the aromatase enzyme system in the ovaries and placenta; estrone is also synthesized from estradiol by 17-hydroxy steroid dehydrogenase in the liver. Estriol is the principal estrogen formed by the placenta during pregnancy. Serum concentrations of estrone in premenopausal women fluctuate according to the menstrual cycle; however, due to the dramatic decline in ovarian estradiol production in menopause, estrone becomes the most predominant estrogen in postmenopausal women.[3] Estrogen action in cells is mediated primarily through ligand-dependent activation of estrogen receptors (ER)-α and -β, although ligand-independent activation of estrogen receptors and nonnuclear actions of estrogens have been described.[3] It is postulated that ERα and ERβ have different or even opposite biological actions, demonstrating a yin-yang relationship.[4] ERβ may negatively regulate cellular proliferation and have a protective role in the normal breast,[5] therefore there is substantial interest in ERβ-selective ligands.[4,6] Based on ligand binding studies, estrone and estriol have lower affinity for both ERs than estradiol,[7] and thus are considered "weaker" estrogens.

The pomegranate is unusual botanically in that it contains relatively high concentrations of *bona fide* naturally occurring estrogens. In fact, the seed of the pomegranate contains the highest known botanical concentration of estrone, reported to be up to 17 mg/kg dried seed.[8] The more common estrogenic compounds of plants, usually referred to as phytoestrogens, are secondary metabolites produced in a wide variety of plants that induce biological responses in vertebrates and can mimic or modulate the actions of endogenous estrogens, usually by binding to estrogen receptors.[9] Phytoestrogens are biphenolic compounds with structures resembling natural and synthetic estrogens (see Figure 9.1). The broad classes of phytoestrogenic compounds include isoflavonoids, coumestans, lignans, and stilbenes.[10,11] Specific classes of isoflavonoids include the flavones (e.g., apigenin), flavonols (e.g., quercetin and kaempferol), flavanones (e.g., naringenin), isoflavones (e.g., genistein, daidzein, and equol), and anthocyanidins; coumestrol is a coumestan and resveratrol, a stilbene.[11–13] Phytoestrogens have the ability to bind the estrogen receptor (ER) due to their biphenolic structure, which is required for the ligand-receptor association, and can act like partial ER agonists or antagonists.[14]

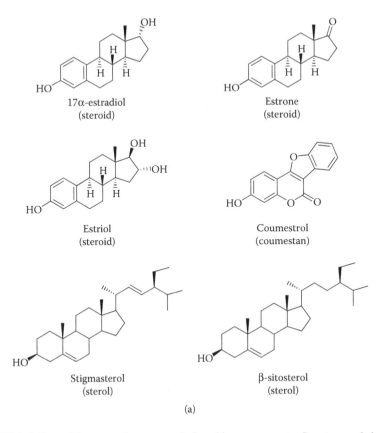

(a)

FIGURE 9.1 Potential estrogenic compounds found in pomegranate. Structure and chemical class of various compounds with potential estrogenic activity found in pomegranate seed (a) and juice/peels (b) are shown.

Many phytoestrogens show a pattern of differential binding to the two ER subtypes, ERα and ERβ, with stronger binding to ERβ as evaluated by radioligand binding assay.[15] The estrogen receptor is part of the steroid/retinoid receptor gene superfamily, a class of soluble DNA binding proteins that act as ligand-activated enhancer factors.[16] Upon ligand binding, the estrogen receptor initiates transcriptional activation by binding to specific palindromic sequences called estrogen response elements (EREs) in the promoters of target genes.[17] The differential effects of agonists and antagonists on receptor activity in a given cell context have been ascribed to different conformations of the receptor–ligand complex, as well as by differences in interaction with transcriptional coactivator and corepressor proteins as well as other transcription factors.[18] In addition, many phytoestrogenic compounds have other cellular activities not ascribed to activation of the ERs, such as regulation of cell signaling pathways, and can have activity in inhibiting proliferation and inducing apoptosis in ER-negative breast cancer cell lines (e.g., MDA-MB-231) as well as in ER-positive lines (such as MCF-7).[19]

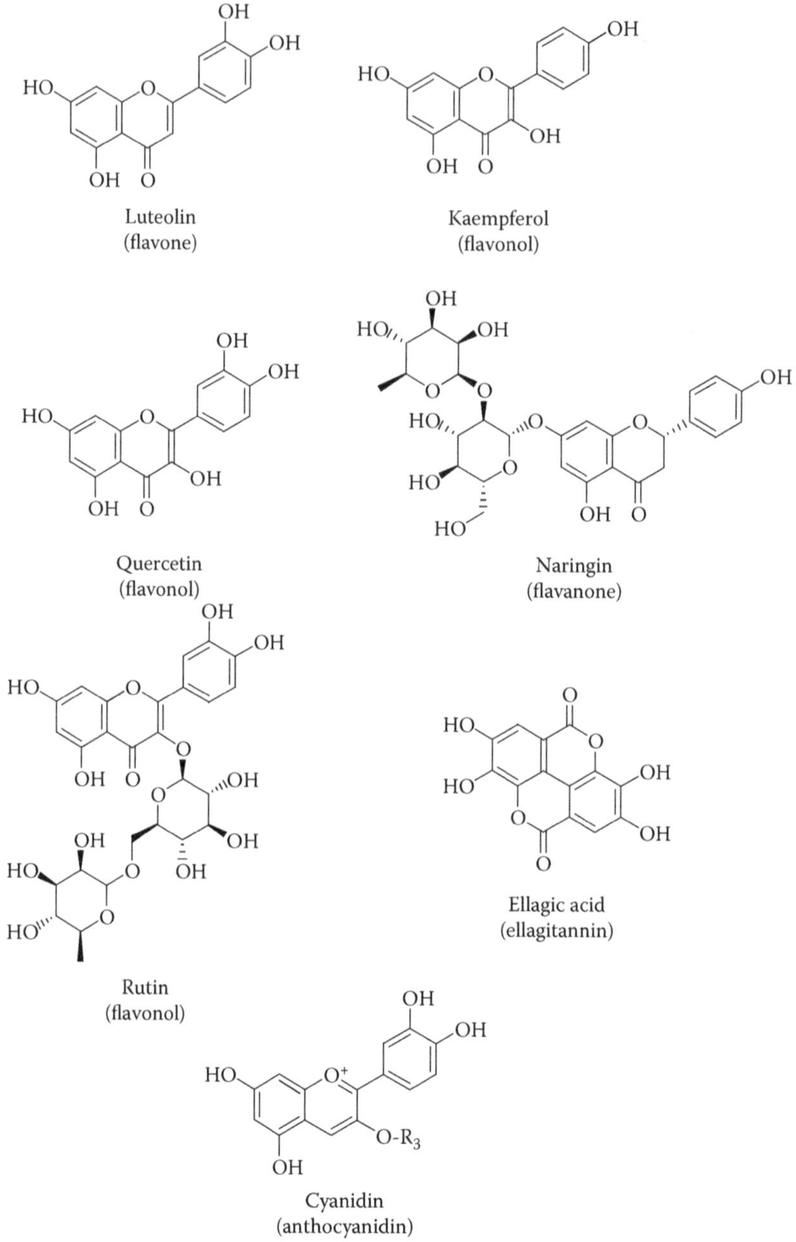

Luteolin
(flavone)

Kaempferol
(flavonol)

Quercetin
(flavonol)

Naringin
(flavanone)

Rutin
(flavonol)

Ellagic acid
(ellagitannin)

Cyanidin
(anthocyanidin)

(b)

FIGURE 9.1 (continued)

There is currently great interest in identifying natural sources of estrogenic agents. Use of pharmacological hormone-replacement therapy for treatment of menopausal symptoms has become controversial, prompting efforts to identify natural sources of estrogens.[20] Some epidemiological data suggest that diets rich in phyto-estrogens protect against breast, prostate, and colon cancer, as well as cardiovascular disease and osteoporosis.[21] Immigration studies have shown that rates of breast cancer in first-generation immigrants, who are primarily eating their native phyto-estrogen-rich diet, are low but increase in the second and subsequent generation with an increasingly Western diet.[22,23] However, the health consequences of phyto-estrogen exposure are perhaps not universally beneficial, and in certain situations, could increase disease risk, particularly with use of dietary supplements that provide exceedingly high levels of phytoestrogens relative to those provided in whole foods.[24] Several potential negative adverse effects have been postulated, including procarci-nogenic effects in peri- and postmenopausal women taking supplements to amelio-rate menopausal symptoms, potential effects on reproduction of natural antiestro-gens, and adverse developmental effects on children prenatally (in the diet of the mother) and in the postnatal diet, as in soy-based infant formulas.[25]

9.2   PRECLINICAL STUDIES OF ESTROGENIC ACTIVITY

Several studies have examined the chemical composition of portions of pomegranate fruit, including juice/pericarp and seeds. Estrone content of pomegranate seeds was first measured by Heftmann et al. at 1.7 mg of estrone per 100 g of pomegranate seeds.[8] An estimate reported by Moneam et al. was similar at 1.09 mg estrone per 100 g seeds.[26] However, another estimate of estrone concentration published by Dean et al. is 4000 times lower at 0.4 mg per 100 g seeds, with no β-estradiol identified.[27] Differences between these various reports are likely due to methodological differ-ences, including extraction procedures and sensitivity of analytical equipment. One of these reports also first identified nonsteroidal phytoestrogens, including coumestrol (0.036 mg per 100 g).[26] Additional unpublished data indicate the presence of 17α-estradiol in the seed oil and narigenin in a 50% solids aqueous peel extract.[28] A more contemporary evaluation of the estrogenic compounds in pomegranate using sensitive online β-estrogen receptor bioassay coupled to mass spectrometry identified luteolin, quercetin, and kaempferol in pomegranate peel extract.[29]

The actual bioactivity of individual phytoestrogen compounds as well as the complex pomegranate extract has been evaluated in a number of model systems. In our own survey of phytoestrogen activity of individual compounds in a transfected breast cancer bioassay, we found coumestrol, kaempferol, and naringenin had pref-erence for activating ERβ relative to ERα, with effective concentrations that elicit half-maximal response (EC$_{50}$'s) of 410, 4,100, and 13,100 times that of the native ligand 17β–estradiol; quercetin had little activity.[30] One of the first reports identifying estrogenic bioactivity of a complex extract of pomegranate was a report from Egypt looking at the estrogenic activity of the seed oil in animal models. A hydrolyzed concentrate of the oil increased both uterine weight in immature rabbits and corni-fication of vaginal epithelium in ovariectomized adult mice, both of which are bioassays of estrogenic activity.[31] In the rabbit assay, the activity of 0.5 mL of the oil was almost the same as that of 10 μg of estradiol, and the effects of estradiol and the seed oil were additive. The estrogenic activity of pomegranate seed oil was confirmed in a uterine weight bioassay in mice, showing that the biological potency of the compound isolated from pomegranate seeds was comparable to that of estrone.[8] A yeast estrogen screen, which evaluates transcriptional activation at ER only, indicates lyophilized fresh pomegranate juice showed 55% inhibition of estrogenic activity of 17β-estradiol, however little direct agonist activity.[28]

Several studies show pomegranate extracts inhibit cell proliferation of breast cancer cells *in vitro*. Fermented juice and pericarp polyphenols inhibited breast cancer cell proliferation, and the degree of inhibition was related to the cell phenotype.[28] Estrogen-dependent MCF-7 breast cancer cells were inhibited to a greater extent than estrogen-independent MB-MDA-231 and normal human breast epithelial cells MCF-10A. In both breast cancer lines, the fermented pomegranate juice polyphenols had about twice the antiproliferative effect of fresh pomegranate juice polyphenols. Pomegranate seed oil also inhibited MCF-7 cellular invasion across a Matrigel membrane, a model for metastasis, and elicited apoptosis in MDA-MB-435 ER-negative metastatic human breast cancer cells. Fermented juice polyphenols inhibited cancerous lesion formation induced by a carcinogen (7,12-dimethylbenz[*a*]anthracene [DMBA]) in a murine mammary organ culture.[28] Further examination of this system isolated a specific fraction of the pomegranate fermented juice polyphenols with highest bioactivity, and also showed that pomegranate seed oil inhibited lesion formation in this system as well.[32]

Here we report on use of an *in vitro* bioassay system to screen pomegranate juice and concentrate for ability to transactivate ER α and -β in a parallel assay. We recently developed a transfected breast cancer cell bioassay to determine estrogen agonist and antagonist activity. In this assay system either ERα or ERβ is expressed, and when activated by ligand, the receptor homodimer binds the ERE, activating transcription of the luciferase reporter gene (see Figure 9.2). This luciferase activity

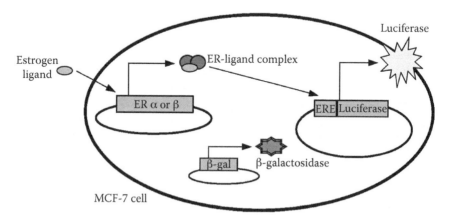

FIGURE 9.2 Principle of the transfection assay used in this study. MCF-7 cells are transiently transfected with one of two ER (ERα or β) expression plasmids plus a luciferase reporter plasmid containing the estrogen response element (ERE). The administered estrogen ligand binds to the estrogen receptor, which forms a homodimer and the receptor-ligand complex now acts as a transcription factor and binds at the ERE site, activating transcription of the luciferase reporter gene. Luciferase activity is assayed and standardized to β-galactosidase.

can be conveniently assayed, giving an indication of activation capacity of one ER relative to the other.

9.3 ASSAY OF ESTROGENIC ACTIVITY OF POMEGRANATE IN A BREAST CANCER CELL BIOASSAY

In the present studies we compared the abilities of commercially available pomegranate juice and concentrate to activate a luciferase gene reporter system in two MCF-7 breast cancer cell lines separately in which either the ERα or ERβ is overexpressed after transient transfection. To investigate additive or potential antagonistic properties we conducted the same experiment in the presence of a constant dose of 17β-estradiol. Pomegranate juice and concentrate were compared at doses from 2 to 512 µg/mL to 17β-estradiol.

9.3.1 EXPERIMENTAL METHODOLOGIES

9.3.1.1 Test Compounds

Pomegranate juice and concentrate (currently not commercially available) were provided by POM Wonderful (Wonderful variety of pomegranate, Los Angeles, CA) and an aliquot of each was freeze-dried overnight and resuspended in 60% EtOH to 1 g/mL final concentration. Samples were stored at –20°C in amber glass vials. The 17β-estradiol was purchased from Sigma Chemical (St. Louis, MO), solubilized in 100% EtOH and stored at –20°C in an amber vial. Prior to the experiment the pomegranate samples as well as the 17β-estradiol were sonicated for 5 minutes.

9.3.1.2 Plasmids

The laboratory of Thomas Scanlan at the University of California, San Francisco, provided plasmids containing either ERα (psG5-HE0), ERβ (psG5-hER), or ERE (EREII-Luciferase GL450) linked to the luciferase reporter gene. The HE0 plasmid contains a point mutation that elicits a lower background response. Use of these plasmids in a similar transfection system has been previously described.[33] We transformed *E. coli* strain DH5α with these plasmids and amplified the plasmids to obtain microgram quantities. An additional β-galactosidase expression plasmid (CMV-β) was cotransfected as an internal standard (Promega, Madison, WI). Plasmid DNA was prepared using a Qiagen Mega-kit (Qiagen, Sunnyvale, CA).

9.3.1.2.1 Cell Culture, Transfection Protocol, and Reporter Gene Assays

The assay was conducted as described previously.[30] In brief, a day before transfection the MCF-7 cells (American Type Culture Collection, Rockville, MD) were plated in 24-well (Corning Incorporated, Corning, NY) plates at a concentration of 8×10^4 cells per well. For transfection, the maintenance media was aspirated and 100 µl transfection mix was added. Transfection media consisted of 0.5 µg DNA divided over the three plasmids (ERα or ERβ, ERE-Luc and CMV-β), 10% heat-inactivated fetal bovine serum, and phenol red-free DMEM/F12 medium (Life Technologies,

Grand Island, NY). Promega's TransFast was used to mediate DNA transfection in the cells. Experimental media was prepared with phenol red-free Dulbecco's modified Eagle's medium/F12, 100 U/mL penicillin, 100 μg/mL streptomycin (Life Technologies, Grand Island, NY), 10% charcoal-stripped fetal bovine serum (HyClone, Logan, UT), and 10 μg/mL insulin (Sigma, St. Louis, MO). For the first experiment, the pomegranate samples were included in four-fold serial dilutions ranging from 2 μg/mL to 512 μg/mL. In the second experiment, cells were incubated with the pomegranate samples at similar doses, but this time in the presence of a constant-level 17β-estradiol (value that gives an average of 80% of the maximal response = 5×10^{-10} M 17β-estradiol (see Figure 9.3). In both experiments, a blank containing 0.2% EtOH was included. Cells were incubated for 48 hours at 37°C, then media was aspirated and cells were lysed with 100 μl of Promega Reporter Gene Lysis Buffer. After a freeze–thaw cycle to ensure lysis, lysates were transferred to microcentrifuge tubes and stored at –80°C. For the luciferase assay, a ratio of 20 μl of lysate per 100 μl of Luciferase Assay System (Promega, Madison, WI) was used. Luminescence was measured with a plate luminometer (Type Berthold Orion, Zylux, TN) with injectors set at a 2-second delay and a 10-second read. For the β-galactosidase activity, a standard protocol was used to assay enzyme activity, using a 20/80 ratio of lysate to reagent (1X galactosidase reagent = 100 mM Na_2HPO_4, 10 mM KCL, 1 mM $MgSO_4$ with substrate [ONPG] added fresh at 4 mg/mL). After a 30-minute incubation, a faint yellow color developed and absorption was measured by a plate-reader at a wavelength of 420 nm. The transfection system described here has been previously shown to have a sensitivity of up to 1 pM for 17-estradiol.

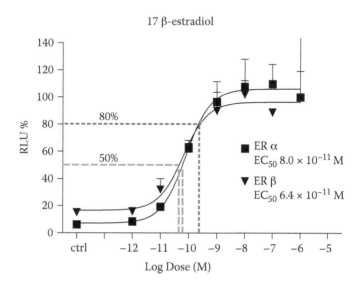

FIGURE 9.3 Estradiol standard curve. EC_{50}'s for 17β-estradiol are calculated for ERα and ERβ (and molar concentration shown on x-axis by dashed lines). The concentration used in the competition experiment, representing 80% of the maximal stimulation, is represented by the dotted lines.

9.3.1.2.2   Statistical Analysis

The software program Prism 3.0 (GraphPad Software Inc., San Diego, CA) was used for data analyses. Light units were divided by the corresponding -galactosidase value and normalized to the maximal value for 17β-estradiol. The EC_{50} values for 17β-estradiol were generated using nonlinear regression analysis. Student's t-test was performed comparing each dose to control (sample with no POM added). Mean values are calculated from triplicate wells averaged over one to two experiments.

9.3.2   RESULTS

Transfected MCF-7 cells were incubated for 48 h with 17β-estradiol to determine the standard curve and with pomegranate juice and concentrate. Production of luciferase enzyme was analyzed to determine estrogen agonist activity in terms of transcriptional activation. The standard curve depicting 17β-estradiol is shown in Figure 9.3. As can be seen, 17β-estradiol elicits a dose-dependent increase in transcriptional activation at both ERα and -β. The EC_{50}'s for stimulation of activity are similar for each ER, and approach the picomolar range. However, as shown in Figure 9.4, the juice and concentrate products did not exhibit estrogenic activity in the dose range evaluated.

To evaluate additive or antagonistic properties of pomegranate with estrogen, serial dilutions of POM juice and POM concentrate were incubated with media containing 5×10^{-10} M 17β-estradiol, a level that represents 80% of the maximal level of transcription (see Figure 9.3). Values were normalized to the sample with 0 μl of pomegranate juice/concentrate, which represents 80% of the maximally achievable response (the constant level of 17ß-estradiol at 5×10^{-10} M) to ascertain if pomegranate is able to compete with the endogenous ligand (17β-estradiol) for the estrogen receptor. There was no significant additive effect of POM with 17β-estradiol in either ER system (Figure 9.5). However, there was a small antagonistic effect of POM concentrate at ERβ for the highest concentrations of POM concentrate.

9.3.3   CONCLUSION AND DISCUSSION

Within the concentration range tested, neither pomegranate juice nor concentrate exhibited an estrogen agonistic response (Figure 9.4). We had hypothesized that these materials may have estrogenic properties, as the pomegranate juice and concentrate tested were obtained by a cold-pressing process of the whole fruit including the seeds, which could result in the extrusion of coumestrol and estrone. In addition, pomegranate is naturally rich in anthocyanins, which have been shown to be able to bind the estrogen receptor with a binding affinity to ERα that is 10,000- to 20,000-fold lower than 17β-estradiol.[34]

There are few previous reports in the literature using similar transfected *in vitro* systems to assay estrogenic effects of pomegranate.[28] One study used yeast *(S. cerevisiae)* instead of mammalian cells. This yeast estrogen screen showed no agonistic effect of pomegranate juice on the ERα subtype of the receptor (ERβ was not evaluated), at a dose of pomegranate juice at 10 μg/mL, in agreement with our results. Just recently, the ability of ellagic acid, a polyphenol abundant in pomegranate

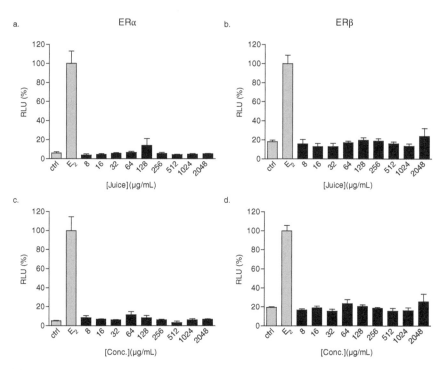

FIGURE 9.4 Effects of pomegranate juice and concentrate on reporter gene transcription in an *in vitro* estrogenicity assay. Transfected MCF-7 cells (see method) were incubated for 48 hours with pomegranate juice (a and b) and pomegranate concentrate (c and d) and levels of luciferase were assayed. Values presented are normalized to the maximally achievable dose (10^{-6} M 17-estradiol, represented as 17-estradiol). Bars presented are SEMs of triplicate values.

pericarp, to independently activate ERα vs. ERβ was determined in a system where HeLa cells were cotransfected with ERα or ERβ expression vectors and ERE-driven luciferase reporter genes.[35] They found ellagic acid could weakly stimulate estrogenic activity at ERα at concentrations between 10^{-7} to 10^{-9} M. However, it acted as an estrogen antagonist at ERβ, a result similar to the antagonistic effect we found in our assay. The authors further examined additional physiological responses. They found that ellagic acid acted as an antiestrogen in MCF-7 cells, increasing insulin-like growth factor binding protein 3 levels at high concentrations (from 10^{-8} to 10^{-6} M), similarly to a pharmaceutical antiestrogen ICI182780 tested at 10^{-8} and 10^{-9} M.

These data suggest that pomegranate concentrate and compounds derived from it may serve as a weak ERβ antagonist *in vitro*. We found that pomegranate concentrate at the two highest doses evaluated inhibits ERβ's activation of a reporter gene. No significant estrogen agonist activity was observed in the estrogen bioassay for either pomegranate juice or pomegranate concentrate. Whether these observations are relevant at doses normally found in the human diet from whole foods, considering issues of bioavailability and maximal intake, remains to be investigated. However,

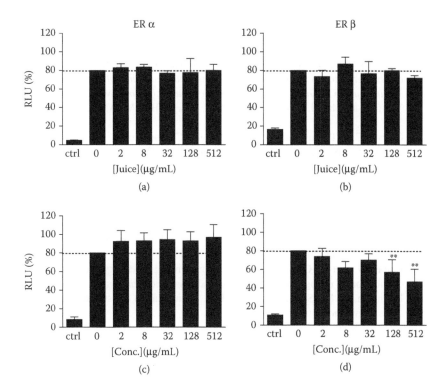

FIGURE 9.5 Effects of pomegranate juice and concentrate on reporter gene transcription in competition with 17β-estradiol. To investigate additive or antagonistic properties of pomegranate, the transfected MCF-7 cells were incubated with serial dilutions of pomegranate juice (a and b) or pomegranate concentrate (c and d) in media containing a constant dose of 5×10^{-10} M 17β-estradiol. Significant inhibition of the level of response was seen with pomegranate concentrate at 128 and 512 μg/L final concentration (**$p < 0.05$). Error bars represent SEMs of two experiments; each sample tested in triplicate.

because of interest in the pomegranate as a food and as a source of raw materials for development of dietary supplements, the effect of potential pomegranate-derived compounds as estrogen receptor modulators should continue to be investigated.

ACKNOWLEDGMENTS

The authors would like to thank M. Koupenova for her technical assistance with the *in vitro* assays. The laboratory of Thomas Scanlan at the University of California, San Francisco, graciously provided the plasmids used in the estrogen bioassay. The Lynda and Stewart Resnick Revocable Trust, Los Angeles, CA, provided funding for the projects described. Additional funding provided by UCLA Center for Dietary Supplement Research: Botanicals (NIH AT00151) and the UCLA Clinical Nutrition Research Unit (NIH CA0A2710).

REFERENCES

1. Langley, P., Why a pomegranate? *BMJ*, 321, 1153, 2000.
2. Lansky, E., Shubert, S., and Neeman, I., Pharmacological and therapeutic properties of pomegranate, in *Production, processing and marketing of pomegranate in the Mediterranean region: Advances in research and technology*, Melgarejo-Moreno, P., Martínez-Nicolás, J.J., Martínez-Tomé, J., Eds., CIHEAM-IAMZ, Zaragoza, 2000, 231.
3. Gruber, C.J., Tschugguel, W., Schneeberger, C., and Huber, J.C., Production and actions of estrogens. *N. Engl. J. Med.*, 346, 340, 2002.
4. Lindberg, M.K. et al., Estrogen receptor (ER)β- reduces ERα-regulated gene transcription, supporting a "Ying Yang" relationship between ERα and ERβ in mice, *Mol. Endocrinol.*, 17, 203, 2003.
5. Hilakivi-Clarke, L. et al., Do estrogens always increase breast cancer risk? *J. Steroid Biochem. Mol. Biol.*, 80, 163, 2002.
6. Gustafsson, J.A., Estrogen receptor beta — a new dimension in estrogen mechanism of action, *J. Endocrinol.*, 163, 379, 1999.
7. Kuiper, G.G. et al., Comparison of the ligand binding specificity and transcript tissue distribution of estrogen receptors alpha and beta, *Endocrinology*, 138, 863, 1997.
8. Heftmann, E., Ko, S.T., and Bennett, R.D., Identification of estrone in pomegranate seeds, *Phytochemistry*, 5, 1337, 1966.
9. Setchell, K.D., Phytoestrogens: the biochemistry, physiology, and implications for human health of soy isoflavones, *Am. J. Clin. Nutr.*, 68, 1333S, 1998.
10. Cornwell, T., Cohick, W., and Raskin, I., Dietary phytoestrogens and health, *Phytochemistry*, 65, 995, 2004.
11. Kurzer, M.S. and Xu, X., Dietary phytoestrogens, *Annu. Rev. Nutr.*, 17, 353, 1997.
12. Ross, J.A. and Kasum, C.M., Dietary flavonoids: bioavailability, metabolic effects, and safety, *Annu. Rev. Nutr.*, 22, 19, 2002.
13. Yang, C.S. et al., Inhibition of carcinogenesis by dietary polyphenolic compounds, *Annu. Rev. Nutr.*, 21, 381, 2001.
14. Leclercq, G. and Heuson, J.C., Physiological and pharmacological effects of estrogens in breast cancer, *Biochim. Biophys. Acta*, 560, 427, 1979.
15. Kuiper, G.G. et al., Interaction of estrogenic chemicals and phytoestrogens with estrogen receptor beta, *Endocrinology*, 139, 4252, 1998.
16. Weihua, Z. et al., Update on estrogen signaling, *FEBS Lett.*, 546, 17, 2003.
17. Klein-Hitpass, L. et al., An estrogen-responsive element derived from the 5′ flanking region of the *Xenopus* vitellogenin A2 gene functions in transfected human cells, *Cell*, 46, 1053, 1986.
18. Benassayag, C., Perrot-Applanat, M., and Ferre, F., Phytoestrogens as modulators of steroid action in target cells, *J. Chromatogr. B Analyt. Technol. Biomed. Life Sci.*, 777, 233, 2002.
19. Limer, J.L. and Speirs, V., Phyto-oestrogens and breast cancer chemoprevention, *Breast Cancer Res.*, 6, 119, 2004.
20. Jordan, V.C., Antiestrogens and selective estrogen receptor modulators as multifunctional medicines. 1. Receptor interactions, *J. Med. Chem.*, 46, 883, 2003.
21. Bingham, S.A. et al., Phyto-oestrogens: where are we now? *Br. J. Nutr.*, 79, 393, 1998.
22. Ziegler, R.G. et al., Migration patterns and breast cancer risk in Asian-American women, *J. Natl. Cancer Inst.*, 85, 1819, 1993.
23. Shimizu, H. et al., Cancers of the prostate and breast among Japanese and white immigrants in Los Angeles County, *Br. J. Cancer*, 63, 963, 1991.

24. Katzenellenbogen, J.A. and Muthyala, R., Interactions of exogenous endocrine active substances with nuclear receptors, *Pure Appl. Chem.,* 75, 1797, 2003.

25. Whitten, P.L. et al., Potential adverse effects of phytoestrogens, *J. Nutr.,* 125, 771S, 1995.

26. Moneam, N.M., el Sharaky, A.S., and Badreldin, M.M., Oestrogen content of pomegranate seeds, *J. Chromatogr.,* 438, 438, 1988.

27. Dean, P.D.G., Exley, D., and Goodwin, T.W., Steroid oestrogens in plants — re-estimation of oestrone in pomegranate seeds, *Phytochemistry,* 10, 2215, 1971.

28. Kim, N.D. et al., Chemopreventive and adjuvant therapeutic potential of pomegranate *(Punica granatum)* for human breast cancer, *Breast Cancer Res. Treat.,* 71, 203, 2002.

29. van Elswijk, D.A. et al., Rapid dereplication of estrogenic compounds in pomegranate *(Punica granatum)* using on-line biochemical detection coupled to mass spectrometry, *Phytochemistry,* 65, 233, 2004.

30. Harris, D.M. et al., Phytoestrogens induce differential estrogen receptor alpha- or beta-mediated responses in transfected breast cancer cells, *Exp. Biol. Med. (Maywood),* 230, 558, 2005.

31. Sharaf, A. and Nigm, S.A.R., The oestrogenic activity of pomegranate seed oil, *J. Endocrinol.,* 29, 91, 1964.

32. Mehta, R. and Lansky, E.P., Breast cancer chemopreventive properties of pomegranate *(Punica granatum)* fruit extracts in a mouse mammary organ culture, *Eur. J. Cancer Prev.,* 13, 345, 2004.

33. Paech, K. et al., Differential ligand activation of estrogen receptors ERα and ERβ at AP1 sites, *Science,* 277, 1508, 1997.

34. Schmitt, E. and Stopper, H., Estrogenic activity of naturally occurring anthocyanidins, *Nutr. Cancer,* 41, 145, 2001.

35. Papoutsi, Z. et al., Evaluation of estrogenic/antiestrogenic activity of ellagic acid via the estrogen receptor subtypes ERα and ERβ, *J. Agric. Food Chem.,* 53, 7715, 2005.

10 Absence of Significant Estrogenic Effects in the Postmenopausal Population

Michelle P. Warren, Eliza Ng, Russalind H. Ramos, and Sari Halpert

CONTENTS

10.1 ESTROGENS, PHYTOESTROGENS, AND THEIR EFFECTS IN WOMEN

Estrogens are categorized according to their structure into two groups: Steroidal and nonsteroidal. Each group contains both natural and synthetic compounds. The steroidal estrogens contain the same basic structure of 18 carbons. The three principal natural steroids — estrone, estradiol, and estriol — are considered the classic estrogens because they were the first to be isolated.[1] Numerous nonsteroidal estrogens have been isolated from natural sources or have been synthesized. Their affinity for the estrogen receptor is very variable.[2]

The phytoestrogens[3] or plant estrogens can show estrogenic effects but are less then 1% as potent as endogenous estrogens. They can have both estrogenic and antiestrogenic properties. The classic steroidal estrogens are not found in plants and are formed in the laboratory from the synthesis of precursors. Plant extracts were first reported to exhibit estrogenic activity in 1926.[4] By 1975, several hundred plants had been found to exhibit estrogenic activity on bioassay or to contain estrogenically active compounds.[5] Phytoestrogens have been identified in bile, urine, semen, blood, and feces in man and animals.

The rapidly growing body of literature on the geographic differences in the incidence and prevalence of many diseases including coronary heart disease; breast, endometrial, and ovarian cancers; and menopausal symptoms, especially hot flushes, have implicated several etiologic factors including racial characteristics, diet, and lifestyle. This has provoked interest in diet, and suggested that certain foods may contain different biologically active compounds.

The consumption of a diet rich in phytoestrogens, commonly seen in Asian populations, especially Japan, have reduced the risk for the so-called "Western" diseases including cardiovascular disease, breast and endometrial cancers, and meno-pausal symptoms such as hot flushes.

The three main classes of phytoestrogens (isoflavones, lignans, and coumestans) occur in either plants or their seeds. Resorcyclic acid lactones exhibit estrogenic activity and are produced by molds that commonly contaminate cereal crops and hence are better termed mycoestrogens. A single plant often contains more than one class of phytoestrogen. For example, the soybean is rich in isoflavones, whereas the soy sprout is a potent source of coumestral, the major coumestan.[6]

Based on their structural similarity to 17β-oestradiol and diethylstilbestrol, with a similar distance between the two –OH groups on equol, enterolactone, and enterodiol to that of 17β-oestradiol, it is thought that phytoestrogens in the diet may modulate hormone-related disorders. These structural configurations are thought to be a factor essential for strong binding to the estrogen receptor. The structure of phytoestrogens is also similar to the weak estrogen/antiestrogen tamoxifen, which is used for the treatment and prevention of breast cancer. Also, the presence and position of the –OH groups on the phytoestrogen compounds, estradiol and dieth-ylstilbestrol, are considered one of the prerequisites for estrogenic activity.[7]

Phytoestrogens are attenuated estrogens.[8] *In vitro* studies performed with human cell cultures have examined the relative estrogenic effects of different phyto-estrogens.[9] Estimates of the relative potencies of these compared to estradiol (E2) at (100%) are coumestrol (0.202), genistein (0.084), equol (0.061), and daidzein (0.013). Importantly, however, the same levels of bioactivity were produced by the isoflavones and by E2 when tested at concentrations sufficiently high to elicit max-imal responses. This indicates that the estrogen receptor complexes formed by E2 and isoflavonoids are functionally equivalent. The comparative dissociation constant of genistein for the estrogen receptor determined in competitive binding experiments is 100 to 10,000 times higher than that of estradiol and diethylstilbestrol.[10] Phyto-estrogens appear to act like "weak estrogens," having less binding affinity for the estrogen receptor as compared to estradiol. Recently, it has been discovered that there are at least two forms of the estrogen receptor, ERα, the "classical" estrogen receptor, and ERβ. It is thought that ERα are found primarily in the breast and uterus, while ERβ are found in the bone and cardiovascular system.[11] However, this is still a subject of investigation and controversy. It is interesting to note that genistein, but not daidzein or glycitein, binds selectively to ERα.[12] In addition, because phytoestrogens exhibit a weaker binding affinity to both estrogen receptors than do endogenous estrogens, they may have much higher concentrations in the body.[13,14] The most common forms of phytoestrogens are the isoflavones; these have a common diphenolic structure resembling the structure of the potent synthetic estrogens diethylstilbestrol and hexestrol. Genistein and daidzein are two of the major isoflavones found in humans. Genistein and daidzein are parent compounds that are metabolized from their plant precursors, biochanin A and formononetin, respectively. Isoflavones are inactive when present in the bound form as glycosides in these plants. These compounds become activated when the sugar residue is removed. Isoflavones are found in a variety of plants, including fruits and vegetables, but they are predominantly found in leguminous plants and are especially abundant in soy.[15] The isoflavone contents of different varieties, crops, and harvest years of soybeans vary substantially. The most extensively studied isoflavone, genistein, is an inhibitor of tyrosine protein kinases,[16] DNA topoisomerases I and II,[17] and ribosomal S6 kinase.[18] Other properties include inhibition of angiogenesis[19] and differentiation of cancer cell lines.[20] Genistein is reported to inhibit tumor promoter-induced hydrogen peroxide formation and to scavenge exogenously added hydrogen peroxide in human cell culture.[21] Other isoflavones such as daidzein, apigenin, and prunectin have also been shown to be potent hydrogen peroxide scavengers and antioxidants.[22] The antioxidant potencies of isoflavones are structurally related and closely associated with the presence of hydroxyl groups at positions 4′ and 5′ and the position of the aromatic ring.[23]

Tyrosine kinase inhibition is important as these receptor enzymes are involved in control of mitogenesis, cell-cycle regulation, cell survival, and cellular transformation via growth factor binding. Growth control factors modulated by tyrosine kinases include epidermal growth factor, transforming growth factor alpha, platelet-derived growth factor, insulin, and insulin-like growth factors, and all have been implicated in tumor growth.[24] Inhibition of DNA topoisomerase II and ribosomal S6 kinase may lead to protein-linked DNA strand breaks, arrest of tumor cell growth, and differentiation induction of several malignant cell lines.[25] Tyrosine kinase mediation of mammary tumor cells to milk-producing, growth-arrested cells has been reported.[26] Possible mechanisms for the antiproliferative properties of genistein include prevention of cell mutations by stabilization of cell DNA and reduction of cell oxidants, reduction in capacity of malignant cells to metastasize by inhibiting angiogenesis and subsequent tumour growth, and induction of cell differentiation.[27] In addition, isoflavones also possess antihypertensive and anti-inflammatory properties.[28]

Lignans possess a 2,3-dibenzylbutane structure and are minor constituents of many plants. They form the building blocks for the formation of lignin (as distinguished from lignan) found in the plant cell wall. They are constituents of higher plants (gymnosperms and angiosperms) such as whole grains, legumes, vegetables, and seeds, with exceptionally high concentrations of lignans found in flaxseed.[29] The chemical structure of plant lignans differs somewhat from that of mammalian lignans. Structural changes of plant lignans occur in the colon, liver, and small intestine during enterohepatic circulation. Mammalian lignans have phenolic hydroxyl groups in the meta position only in their aromatic rings, differing from plant lignans. Once in the colon, they are absorbed and then are conjugated with glucuronic acid or sulfate in the liver, reexcreted through the bile duct, deconjugated by the bacteria, and reabsorbed. Some reach the kidney and are excreted in the urine.[30] Lignans are excreted in the urine as conjugated glucuronides and in feces in the unconjugated form.[31]

Many plant lignans have been shown to have anticarcinogenic, antiviral, bactericidic, and fungistatic activities.[32–34] Enterolactone, the most abundant mammalian lignan, is a moderate inhibitor of placental aromatase and competes with the natural substrate androstenedione for the enzyme. Other experiments with a choriocarcinoma cell line (JEG-3) showed that enterolactone is very readily transferred from cell culture media into the cells.[35] Most of the lignans, as well as flavonoids, are only weak inhibitors. However, a diet rich in vegetables may, due to the abundance of these compounds in the diet, lead to sufficient concentrations (e.g., in fat cells) to reduce conversion of androstenedione to estrone, lowering risk for estrogen-dependent cancer.[36] Lignans may also affect cholesterol homeostasis, as they have been shown to inhibit the activity of cholesterol-7 hydroxylase, the rate-limiting enzyme in the formation of primary bile acids from cholesterol.[37]

10.1.1 HOT FLUSHES[38]

Asian women complain less about hot flushes (also called hot flashes) and other menopausal symptoms than do Western women. Although it is difficult to factor in the effect of cultural differences (e.g., the fact that Asians respect age while Westerners worship youth), it has been suggested that eating soy products serves as a type of hormone replacement therapy.[39]

There is evidence that phytoestrogen supplementation can help hot flashes, although studies are mixed. A double-blind placebo-controlled study randomized 104 postmenopausal women to 60 g of isolated soy protein or to casein as a control for 12 weeks. Hot flushes were reduced by 45% in the soy group compared with 30% in the placebo group (the difference was statistically significant).[40] Differences between the two groups were evident by the third week. Adverse effects (mainly gastrointestinal complaints) were similar in the two groups.

A 12-week study in 145 Israeli women with menopausal symptoms (114 completed the study) found that the 78 women assigned to a phytoestrogen-rich diet (approximately one-fourth of caloric intake) had fewer hot flushes than the 36 women in the control group.[41] Total scores on a menopausal symptom questionnaire were not significantly different between the two groups.

Another study with 58 women enrolled compared daily intake of 45 g of soy flour with the same amount of wheat flour for 12 weeks. The soy-treated group has significantly fewer hot flushes.[42] Other studies have found soy to have no effects on alleviation of postmenopausal symptoms.[43,44]

Pomegranate juice has been found by a variety of different bioassays to have some estrogenicity. In particular, the seed oil has been found to increase uterine weight and cause a dose-dependent effect on the cornification of vaginal smears in rabbits when compared to controls.[45] This was not confirmed by two other studies on the uterine weights of rabbits using juice and press by-product.[46] The seeds were found to contain varying levels of estrone: 8 mg/kg,[47] 17 mg/kg,[48] 10 mg/kg,[49] and 0.004 mg/kg.[50] A study of seeds conducted by Better Living Japan suggests the presence of the phytoestrogen coumestrol at 0.017 to 0.76 mg/kg, while none of the concentrates do. Mass spectrometry has not confirmed the presence of estrone or phytoestrogens belonging to the isoflavone class.[51] Plants do not contain steroidal estrogens but generally, if used as a source of steroid synthesis, provide precursors that are synthesized in the laboratory to estrogenic steroids. It would be highly unusual to find sex steroids in plants, and humans cannot synthesize these steroids from plant precursors. Estrogenic compounds found in plants are generally phytoestrogens. Any increase in estrogen levels therefore would be through suppressing a metabolic pathway that inhibits estrogen synthesis *(vide infra)*. However, there are some exceptions,[48,49] which suggest that steroidal estrogen is made in pomegranate seeds and in date palm.

Some plants may influence the metabolic pathways of endogenous sex steroids in particular by influencing the enzyme cytochrome P450 aromatase, which is present in many tissues including adipose tissue and is responsible for transforming androstenedione into estrone. The major source of estrone in the menopausal women is peripheral aromatization of androstenedione. Ninety-five percent of this hormone comes from the adrenal gland and only 5% from the ovary. The conversion is increased in obesity.

The average level of estrone in postmenopausal women is 30 pg/ml ± 11 pg/ml. The levels of estrone in pregnant women is 1,183 to 12,230 pg/ml, and levels in men are 41.3 pg/ml.

We performed a study to determine if pomegranate juice elevated levels of any of the three classic estrogens, estadiol, estriol, and estrone, or had an effect on vaginal maturation, an indicator of physiological effects, when ingested by postmenopausal women.

10.2 A STUDY OF ESTROGENIC EFFECTS IN THE POSTMENOPAUSAL POPULATION

10.2.1 Design and Procedures

In this study, 13 postmenopausal women were screened and 11 were enrolled (Table 10.1). The enrolled patients were 52–63 years old, had a mean age of 57 years, were 2–12 years postmenopausal, and included one patient who underwent a hysterectomy 23 years ago. The BMI ranged from 17 to 34 with an average of 26.4, two patients

TABLE 10.1
Characteristics of Study Participants

Description	n	Baseline Mean ± SD
Age	11	57.18 ± 3.74
Years since Menopause	11	7.95 ± 6.37
Weight	11	150.5 ± 26.36
Height	11	160.35 ± 6.48
BMI	11	26.42 ± 4.26

were smokers, three patients reported a history of high cholesterol, and one reported hypothyroidism. Most of the patients were taking multivitamins and calcium supplements.

This study involved three visits. On visit 1, baseline levels of follicle stimulating hormone (FSH), estradiol, estrone, and estriol were measured after the patients signed an informed consent. On visit 2, medical history, physical exam, and pelvic exam including a vaginal smear to assess degree of vaginal cornification were done and were uneventful except for one patient who had a benign cervical polyp and underwent a polypectomy. All patients had a negative allergy test to pomegranate juice. Patients were then given a 1-week supply of 8 oz of pomegranate juice (Wonderful variety, POM Wonderful, Los Angeles, CA), a patient instruction form on juice intake, and a diary to report juice compliance and any adverse events. All subjects reported daily intake of pomegranate juice for 7 days with no adverse events. On visit 3, which was scheduled on the seventh day of juice intake, levels of FSH, estradiol, estrone, and estriol were measured to assess post-test hormonal profile. Patients had a physical exam and pelvic exam including a post-test vaginal smear. One subject had a baseline and post-test vaginal smear report of inflammation, and was therefore not included in the data analysis of vaginal maturation index.

10.2.2 Results and Discussion

The null hypothesis was that there is no difference in the hormonal profiles and vaginal smear maturation index before and after the pomegranate juice ingestion. Two-tailed t-tests were used to compare the levels of estradiol, estrone, and FSH at visit 1 and visit 3. All but one subject exhibited a higher level of estrone at visit 3, with a mean group difference of -23 ng/ml, which is statistically significant (95% CI $= -35.8$ ng/ml to -11.627 ng/ml; two-tailed paired t-test, $t = -4.369$; $p < 0.001$) (Table 10.2). There were no statistical differences in the levels of estradiol and FSH between visit 1 and visit 3. On average, the level of estradiol was 0.182 higher in visit 3, and the FSH was 0.273 lower on visit 3 (Table 10.2). The level of estriol in all subjects was undetectable both at baseline and at visit 3. There was also no difference in the maturation index between visit 1 and visit 3 (Table 10.3). The level of superficial, intermediate, and parabasal cells detected by the vaginal smear

TABLE 10.2
Comparison of Study Participant Hormone Levels at Baseline and Follow-Up

Hormones	n	Baseline Mean ± SD	Day 7 Mean ± SD	P
Estradiol	11	26.73 ± 4.36	26.55 ± 8.79	NS
Estrone	11	40.82 ± 10.18	64.55 ± 15.83	0.001 Significant
Estriol	11	<0.1	<0.1	NS
FSH	11	70.82 ± 17.59	71.09 ± 24.31	NS

TABLE 10.3
Comparison of Vaginal Smears between Baseline and Follow-Up

Vaginal Smear	n	Baseline Mean ± SD	Day 7 Mean ± SD	P
Intermediate	10	62.50 ± 24.30	51.50 ± 30.65	NS
Parabasal	10	57.50 ± 31.29	46.50 ± 34.72	NS

were again analyzed using paired t-tests, and none showed a statistical difference (Table 10.3).

This pilot study showed no increase in the levels of estradiol or estriol and no significant biologic estrogen effect on vaginal cornification, as demonstrated by the lack of increase in superficial cells on vaginal smear. There was also no change in the pituitary gonadotropins, LH and FSH, demonstrating no significant estrogenic feedback effect. However, there was a significant increase in estrone of 40.82 pg/ml to 64.55 pg/ml.

The average level of estrone in postmenopausal women is 30 pg/ml ± 11 pg/ml. Our power calculations, with a minimal detectable difference of 20 pg/ml, indicated that we need a sample size of five subjects (see study design and statistical analysis). Thus, this appears to be significant as the increase was more than 20 pg/ml and should initiate a more detailed study.

However, despite this increase, estrone would not be expected to have estrogenic effects, as a physiologic effect would be expected only at much higher levels (i.e., 150 to 300 pg/ml). Furthermore, estrone is less potent than estradiol (50 to 70% as potent) and the only metabolite it produces that has estrogenic activity is estriol, which was not significantly increased. In fact, no statistical analyses were performed on estriol, as it remained below the detection limit of the assay for the study, including after 7 days ingestion of pomegranate juice.

The major source of estrone in the menopausal women is peripheral aromatization of androstenedione. Ninety-five percent of this hormone comes from the adrenal gland and only 5% from the ovary. The conversion is increased in obesity. While some subjects were obese (three subjects with Body Mass Index (BMI kg/m^2 > 30), the increase in estrone was not related to BMI. The two subjects with some superficial cells on vaginal smear at baseline (superficial cells are a reflection of an estrogenic effect) had BMIs of 27.9 (50% superficial cells) and 30.4 (25% superficial cells). However, the superficial cells decreased in the latter subject after pomegranate juice, and overall there was no significant difference in the number of superficial cells after supplementation of pomegranate juice for 7 days.

There was no relationship between the change in estrone and age. However, there was an inverse relationship to years since menopause, with the subjects closer to the age of menopause having the largest increase in estrone. The only baseline level that predicted the poststudy level was the vaginal smear in terms of superficial cells. No other value was predicted by the baseline level (see paired samples correlations).

Thus, in conclusion, there appears to be a significant increase in estrone at a nonphysiologic level in postmenopausal women after a week of pomegranate juice supplementation. Whether there would be a continued cumulative effect with long-term use is unclear. Since plants generally do not contain steroidal estrogens (with certain exceptions), we conclude that if estrone levels are higher in women drinking pomegranate juice, there may be a mild effect on the enzyme cytochrome P450 aromatase activity.

REFERENCES

1. Stanczyk, F.Z., Estrogens: different types of properties, in *Menopause: biology and pathology,* Lobo, R.A., Kelsey, J., and Marcus, R., eds., Academic Press, San Diego, 2000, chap. 29.
2. Stanczyk, F.Z., Estrogens: different types of properties, in *Menopause: biology and pathology,* Lobo, R.A., Kelsey, J., and Marcus, R., eds., Academic Press, San Diego, 2000, chap. 29.
3. This section taken from Warren, M.P., Shortle, B., and Dominguez, J.E., Use of alternative therapies in menopause, *Best Pract. Res. Ob.,* 16, 411, 2002.
4. Loewe, S.R., Lange, F., and Spohr, E., Uber weiliche sexual hormone (thelytropine), *Biochem. Zeitschr.,* 180, 1, 1927.
5. Fransworth, N.R. et al., Potential value of plants as sources of new antifertility agents II, *J. Pharm. Sci.,* 64, 717, 1975.
6. Price, K.R. and Fenwick, G.R., Naturally occurring oestrogens in foods — a review, *Food Addit. Contam.,* 2, 73, 1985.
7. Martucci, C.P. and Fishman, J., Enzymes of estrogen metabolism, *Pharmacol. Ther.,* 57, 237, 1993.
8. Bannwart, C. et al., Isoflavonic phyto-estrogens in humans, identification and metabolism, *Eur. J. Cancer Clin. On.,* 23, 1732, 1987.
9. Markiewiez, L. et al., *In vitro* bioassays of non-steroidal phytoestrogens, *J. Steroid. Biochem. Molec. Biol.,* 45, 399, 1993.
10. Martin, P.M. et al., Phytoestrogen interaction with estrogen receptors in human breast cancer cells, *Endocrinology,* 103, 1860, 1978.
11. Couse, J.F. et al., Tissue distribution and quantitative analysis of estrogen receptor-alpha (ERα) and estrogen receptor-beta (ERβ) messenger ribonucleic acid in the wild-type and ERα-knockout mouse, *Endocrinology,* 138, 4613, 1997.
12. Peterson, G. and Barnes, S., Genistein inhibition of the growth of human breast cancer cells: independence from estrogen receptors and the multi-drug resistance gene, *Biochem. Biophys. Res. Commun.,* 179, 661, 1991.
13. Tham, D., Gardner, C., and Haskell, W., Potential health benefits of dietary phyto-estrogens: a review of the clinical, epidemiological, and mechanistic evidence, *J. Clin. Endocrinol. Metab.,* 83, 2223, 1998.
14. Adlercreutz, H. and Mazur, W., Phyto-oestrogens and Western diseases, *Ann. Med.,* 29, 95, 1997.
15. Tham, D., Gardner, C., and Haskell, W., Potential health benefits of dietary phyto-estrogens: a review of the clinical, epidemiological, and mechanistic evidence, *J. Clin. Endocr. Metab.,* 83, 2223, 1998.
16. Akiyama, T. et al., A specific inhibitor of tyrosine-specific protein kinase, *J. Biol. Chem.,* 262, 5592, 1987.

17. Okura, A. et al., Effect of genistein on topoisomerase activity and on the growth of [val 12] Ha-ras-transformed NIH 3T3 cells, *Biochem. Biophys. Res. Commun.*, 157, 183, 1988.

18. Linassier, C., Pierre, M., Le Pecq, J.B., and Pierre, J., Mechanisms of action in NIH-3T3 cells of genistein, an inhibitor of EGF receptor tyrosine kinase activity, *Biochem. Pharmacol.*, 39, 187, 1990.

19. Fotsis, T. et al., Genistein, a dietary derived inhibitor of *in-vitro* angiogenesis, *Proc. Nat. Acad. Sci. USA*, 190, 2690, 1993.

20. Constantinou, A. and Huberman, E., Genistein as an inducer of tumor cell differentiation: possible mechanisms of action, *Proc. Soc. Exp. Biol. Med.*, 208, 109,1995.

21. Wei, H. et al., Inhibition of tumour promoter-induced hydrogen peroxide formation *in vitro* and *in vivo* by genistein, *Nutr. Cancer*, 20, 1, 1993.

22. Wei, H. et al., Antioxident and antipromotional effects of the soybean isoflavone genistein, *Proc. Soc. Exp. Biol. Med.*, 208, 124, 1995.

23. Tham, D., Gardner, C., and Haskell, W., Potential health benefits of dietary phytoestrogens: a review of the clinical, epidemiological, and mechanistic evidence, *J. Clin. Endocr. Metab.*, 83, 2223, 1998.

24. Knight, D.C. and Eden, J.A., Phytoestrogens — a short review, *Maturitas*, 22, 167, 1995.

25. Petersen, T.G. and Barnes, S., Genistein inhibition of the growth of human breast cancer cells: independence from estrogen receptors and the multi-drug resistance gene, *Biochem. Biophys. Res. Commun.*, 179, 661, 1991.

26. Wen, D. et al., Neu differentiation factor: a transmembrane glycoprotein containing an EGF domain and an immunoglobulin homology unit, *Cell*, 69, 559, 1992.

27. Knight, D.C. and Eden, J.A., A review of the clinical effects of phytoestrogens, *Obstet. Gynecol.*, 87, 897, 1996.

28. Knight, D.C. and Eden, J.A., Phytoestrogens — a short review, *Maturitas*, 22, 167, 1995.

29. Tham, D., Gardner, C., and Haskell, W., Potential health benefits of dietary phytoestrogens: a review of the clinical, epidemiological, and mechanistic evidence, *J. Clin. Endocr. Metab.*, 83, 2223, 1998.

30. Hutchins, A.M. et al., Vegetables, fruits, and legumes: effect on urinary isoflavonoid phytoestrogen and lignan excretion, *J. Am. Diet Assoc.*, 95, 769, 1995.

31. Tham, D., Gardner, C., and Haskell, W., Potential health benefits of dietary phytoestrogens: a review of the clinical, epidemiological, and mechanistic evidence. *J. Clin. Endocr. Metab.*, 83, 2223, 1998.

32. Setchell, K.D.R. and Adlercreutz, H., Mammalian lignans and phytoestrogens. Recent studies on their formation, metabolism and biological role in health and disease, in Rowland, I., ed., *Role of the gut flora in toxicity and cancer*, London, Academic Press, 1988, 315–45.

33. Rao, C.B.S, ed., *The chemistry of lignans*, 377, 1978, Waltair, India, Andhra University Press.

34. Ayres, D.C. and Loike, J.D., Chemical, biological and clinical properties, in Phillipson, J.D., Ayres, D.C., and Baxter, H., eds., *Chemistry and pharmacology of natural products*, Cambridge, UK, Cambridge University Press, 1990, 402.

35. Adlercreutz, H. et al., Inhibition of human aromatase by mammalian lignans and isoflavonoid phytoestrogens, *J. Steroid. Biochem. Mol. Bio.*, 44, 147, 1993.

36. Henderson, B.E., Ross, R., and Bernstein, L., Estrogens as a cause of human cancer: the Richard and Linda Rosenthal Foundation award lecture, *Cancer Res.*, 48, 246, 1998.

37. Hirose, N. et al., Inhibition of cholesterol absorption and synthesis in rats by sesamin, *J. Lipid Res.,* 32, 629, 1999.

38. Fugh-Berman, A., Complementary medicine: herbs, phytoestrogens, and other treatments, in *Treatment of the postmenopausal woman: basic and clinical aspects,* second edition, Lobo, R.A., ed., Lippincott Williams & Wilkins, Philadelphia, ch. 41, 1999.

39. Aldlercreutz, H. et al., Dietary phytoestrogens and the menopause in Japan, *Lancet,* 339, 1233, 1992.

40. Albertazzi, P. et al., The effect of dietary soy supplementation on hot flushes, *Obstet. Gynecol.,* 91, 6, 1998.

41. Brzezinski, A. et al., Short-term effects of phytoestrogen rich diet on postmenopausal women, *Menopause,* 3, 89, 1997.

42. Murkies, A.L., Lombard, C., and Strauss, B.J.G., Dietary flour supplementation decreases postmenopausal hot flushes: effect of soy and wheat, *Maturitas,* 21, 189, 1995.

43. Quella, S.K. et al., Evaluation of soy phytoestrogens for the treatment of hot flashes in breast cancer survivors: a North Central Cancer Treatment Group trial. *J. Clin. Oncol.,* 18, 1068, 2000.

44. St. Germain, A. et al., Isoflavone-rich or isoflavone-poor soy protein does not reduce menopausal symptoms during 24 weeks of treatment. *Menopause,* 8, 17, 2001.

45. Sharaf, A. and Nigm, S.A.R., The oestrogenic activity of pomegranate seed oil, *J. Endocrinol.,* 29, 91, 1964.

46. Dornfeld, L., personal communication, 2000.

47. Dornfeld, L., personal communication with Ameritech Laboratories, 1999.

48. Heftman, E., Ko, S.T., and Bennet, R., Identification of estrone in pomegranate seeds, 1337–41, 1996.

49. Moneam, N.M.A., Sharaky, A.S.E.L., and Balreldin, M.M., Oestrogen content of pomegranate seeds, *J. Chromatogr.,* 438, 1998.

50. Dean, P.D.G., Exley, D., and Goodwin, T.W., Steroid oestrogens in plants: re-estimation of oesterone in pomegranate seeds, *Phytochemistry,* 10, 2215, 1971.

51. Dornfeld L., personal communication, 2000.

11 Antimicrobial Activities of Pomegranate

G.K. Jayaprakasha, P.S. Negi, and B.S. Jena

CONTENTS

11.1 INTRODUCTION

The present scenario of emergence of multiple drug resistance to human pathogenic organisms has initiated a search for new antimicrobial substances from plant sources. The use of higher plants and preparations made from them to treat infections is an age-old practice in a large part of the world population, especially in developing countries. The pomegranate tree, *Punica granatum* L. (Punicaceae), is a native shrub of occidental Asia and Mediterranean Europe that has a rich history of traditional use in medicine. For centuries, the bark, leaves, flowers, fruits, and seeds of this plant have been used to ameliorate diseases. Several studies have reported the efficacy of various extracts or pure compounds from the different parts of pomegranate plant against the growth of Gram-positive and Gram-negative bacteria. Moreover, the extracts from the pomegranate peel have been found to be active against methicillin-resistant *Staphylococcus aureus* and multidrug-resistant *Salmonella typhimurium*. Furthermore, the extracts from pomegranate were effective against both food-borne pathogens and spoilage bacteria. The antifungal activity of the extracts was prominent against *Candida* sp., and antiviral activities against herpes virus, polyhedrosis virus, poliovirus, human respiratory syncytial virus, and human immunodeficiency virus were also observed. The extracts from pomegranate also showed promising activities as antidiarrheal and tanicidal. The presence of phyto-chemicals such as phenols, tannins, flavonoids, and alkaloids in the pomegranate as major active constituents may be responsible for these biological activities. Hence, there is promise for using pomegranate extracts or purified compounds as a thera-peutic agent and also in food biopreservation.

11.2 MEDICINAL USES OF POMEGRANATES

Punica granatum L. (Punicaceae) is a native shrub of occidental Asia and Mediter-ranean Europe, popularly referred to in English as pomegranate.[1] Ancient Egyptian culture regarded the pomegranate fruit as a symbol of prosperity and ambition, making it common practice to decorate sarcophagi with the plant. According to Ebers' papyrus (one of the oldest medical writings, *circa* 1500 B.C.), the plant was used by Egyptians as a treatment for tapeworm and other parasitic infestations. In ancient Greek mythology, pomegranates are known as the "fruit of the dead." The Babylonians regarded the seeds as an agent of resurrection, the Persians as conferring invincibility on the battlefield, and for ancient Chinese alchemical experts, the bright red juice was mythopoetically regarded as a "soul concentrate," homologous to human blood, and capable of conferring on a person longevity or even immortality.[2] The pomegranate tree, which is said to have flourished in the Garden of Eden, has also been extensively used as a folk medicine in many cultures, documented at least as far back as the Egyptian Papyrus of Ebers, *circa* 1550 B.C.[3] Several parts of this tree have been used as an astringent, haemostatic, as a remedy for diabetes, as an anthelmintic specifically against tapeworms, and for diarrhea and dysentery.[4,5] For centuries the bark, leaves, flowers, and fruit of this plant have been used to ameliorate diseases ranging from conjunctivitis to hematuria.[6] In Brazil the fruits are known as "roma" and are used for the treatment of throat infections, coughs, and fever. There are several commercial phytopreparations in Brazil containing extracts from pome-granate.[7] In India, almost all parts of this plant are used in traditional medicine for the treatment of various ailments. Bark and rind of the fruit are used in dysentery, diarrhea, piles, bronchitis, bilious affection, and as an anthelmintic. The flower bud is used in chronic diarrhea and dysentery of children.[8,9] Pomegranate peel has been used in the preparation of hemostatic sterilizing agent and was found to be an effective disinfectant and helpful in elimination of pus, and also used in preparation of wound-healing material.[10] Seeds of pomegranate are a rich source of crude fiber, sugar, and pectin. They contain glycerol and sterols, and in particular seeds are reported to contain a variable amount of the steroid estrogen estrone,[11] used for a variety of chemotherapeutic purposes.

Finding healing powers in plants is an ancient idea. There is evidence that Neanderthals living 60,000 years ago in present-day Iraq used plants such as holly-hock.[12,13] Historically, plants have provided a source of inspiration for novel drug compounds, as plant-derived medicines have made large contributions to human health. Their role is twofold in the development of new drugs: they become the base for the development of a medicine and a phytomedicine to be used for the treatment of disease. Many commercially proven drugs used in modern medicine were initially used in crude form in traditional or folk healing practices or for other purposes that suggested their potentially useful biological activities.[14] Many plants have been used in traditional medicine because of their antimicrobial traits, which are due to compounds synthesized in the secondary metabolism of the plant. These products are known by their active substances, such as phenolic compounds that are part of the essential oils,[15] as well as in tannin.[16] In recent decades, pomegranate has been studied for many potential uses including immunomodulation, atherosclerosis/arteriosclerosis, angina, skin injuries, urinary tract inflammation, scabies, abscesses, cough, bacterial infection, fungal infection, parasitic infection, periodontal disease, and food poisoning.[17–22] The most famous usage worldwide has been as a vermifugal or tanicidal agent[23,24] (i.e., a killer and expeller of intestinal worms).

The stability of some foods against microbial contamination is due to the fact that they contain naturally occurring substances with antimicrobial activity. The major limitations in the use of naturally derived preservatives are due to associated flavors, which can alter the taste of food. Therefore, the mechanism of growth inhibition by these natural preservatives can lead to new technologies for their use in maintaining the quality of foods. The use of natural antimicrobial compounds is important not only in the preservation of food, but also in the control of human and plant diseases of microbial origin. Bacterial and fungal infections pose a greater threat to health, most notably in immune-compromised subjects.[25] Therefore, the need for naturally derived compounds and other natural products with antimicrobial properties has been explored.[26–31]

11.3 ANTIBACTERIAL ACTIVITY

The use of higher plants and preparations made from them to treat infections is an age-old practice in a large part of the world population, especially in developing countries, where there is dependence on traditional medicine for a variety of diseases.[32] Traditionally used medicinal plants produce a variety of compounds of known therapeutic properties.[33–35] The substances that can either inhibit the growth of pathogens or kill them and have no or least toxicity to host cells can be considered good candidates for developing new antimicrobial drugs. In recent years, antimicrobial properties of medicinal plants are being increasingly reported from different parts of the world.[36–41] Interest in plants with antimicrobial properties has revived as a consequence of current problems associated with the use of antibiotics.[42,43] Infections caused by methicillin-resistant *Staphylococcus aureus* (MRSA) isolates have increased greatly during the last decades in hospitals.[44–46]

There are reports of antimicrobial activity of pomegranate[47–50] showing that pomegranate juice is inhibitory to *Staphylococcus epidermidis* and *Klebsiella pneumoniae*. Antimicrobial activities of different pomegranate extracts extracted from different parts of pomegranate and solvents extracts have been presented in Table 11.1. Kirilenko et al.[51] reported that the antibacterial action of pomegranate juice varied with variety and depended on the contents of phenolic compounds, pigments, and citric acid. De et al.[52] also reported potential antimicrobial activity of pomegranate seeds against *Bacillus subtilis, Escherichia coli,* and *Saccharomyces cerevisiae.*

TABLE 11.1
Antibacterial and Antifungal Activities of Extracts from Pomegranate Fruit Parts against Different Microorganisms

Fruit Parts	Solvent Used for the Extraction/Compound	Microorganism Tested	Reference
Hulls	—	*Salmonella typhi* and *Vibrio cholerae*	55, 56
Fruits	MeOH	*S. aureus* FRI722	6
	—	Candida mycoderma	51
	MeOH, aqueous	*Escherichia coli* O157:H7, *E. coli* O26:H11, *E. coli* O111: NM, *E. coli* O22	61
	Ethanol	Staphylococcus aureus	63
	Petroleum ether, chloroform, MeOH, water	*S. aureus* MTCC 737, *E. coli* MTCC723, *Klebsiella pneumoniae* MTCC 109, *Proteus vulgaris* MTCC 1771, *B. subtilis* MTCC 441, *Salmonella typhi* MTCC 537	21
	EtOH	C. albicans	67
Juice	—	Antibacterial activity	51
Peels	Punicalagin	M. tuberculosis	5
	Ethyl acetate, acetone, MeOH, water	*B. cereus, B. coagulens, B. subtilis, Staphylococcus aureus, E. coli,* and *Pseudomonas aeruginosa*	30, 53
	Punicalagin	Candida albicans	53
	MeOH, aqueous	*Salmonella typhi* (MTCC 531, B 330)	57
	EtOH	Ralstonia solanacearum	68
	—	*P. citrinum, P. patulum; P. roquefortii, A. ochraceus, A. flavus,* and *A. parasiticus*	69
Pericarp	Ethanol/Punicalin	Staphylococcus aureus	7
	MeOH	*S. aureus, Eschericia coli, Pseudomonas aeruginosa,* and *Candida albicans*	54
	Hexane, chloroform, dichloromethane, ethyl acetate, butanol, water/Punicalagin	Six methicillin-resistant stains of *S. aureus*	64
Seeds	Aqueous	*B. subtilis, Escherichia coli,* and *Saccharomyces cerevisiae*	52

Pomegranate fruit peel compound punicalagin is reported to have antimicrobial activity against *Staphylococcus aureus* and *Pseudomonas aeruginosa*.[53] Prashanth et al.[21] reported antibacterial activity of petroleum ether, chloroform, methanol, and water extracts of pomegranate rinds. However, the methanolic extract was found to be most effective against all tested microorganisms such as *Staphylococcus aureus, Escherichia coli, Klebsiella pneumoniae, Proteus vulgaris, Bacillus subtilis,* and *Salmonella typhi*. The phytochemical screening gave positive tests for sterols for petrol extract, alkaloids for chloroform, tannins and alkaloids for methanol extract, and tannins for water extract.[21] Negi and Jayaprakasha[30] extracted pomegranate peels with different polar solvents at room temperature and assayed them for antibacterial activity. Acetone, methanol, and water extracts were evaluated against both Gram-positive and Gram-negative bacteria. The acetone extract showed the highest antibacterial activity, followed by methanol and water extract. Methanolic extract from the fruit pericarp of pomegranate was found to be antimcrobial against *S. aureus, E. coli, Pseudomonas aeruginosa,* and *Candida albicans.*[54] Studies have shown that the simple extracts of the hulls have efficacy against the virulent intestinal bacteria *Salmonella typhi*[55] and *Vibrio cholerae.*[56] Rani and Khullar[57] have reported that both aqueous and methanol extracts from the peel of pomegranate showed strong antibacterial activity against multidrug-resistant *Salmonella typhi.* However, methanol extract showed higher activity than aqueous extract. The results of this study are corroborated with the results of previous studies[21,54,58] in which these plant extracts were used as antimicrobial agents and the methanol extract was found to be more effective. The authors also inferred that the presence of ellagitannins, gallotannins,[59] and alkaloids[60] in pomegranate might be responsible for antimicrobial activity.

Aqueous and ethanolic fruit shell extracts of pomegranate were found to have antibacterial activity against different strains of *Escherichia coli,* including six strains of *E. coli* (Table 11.1); five strains of *E. coli* isolated from bovine; and *E. coli* ATCC 25922.[61] However, the aqueous extract of pomegranate was highly effective against enterohemorrhagic *E. coli* O157:H7. This plant species may provide alternative bioactive medicines for the treatment of *E. coli* O157:H7 infection. Various extracts from pomegranate showed antibacterial activity against enterohemorrhagic *E. coli.*[21,38,62]

Studies have shown that pomegranate is effective in inhibiting the growth of *Staphylococcus aureus.*[63] The ethanol extract of the pericarp of pomegranate suspended in water and successively partitioned with hexane, chloroform, ethyl acetate, and butanol. Ethyl acetate, the most active fraction, was subjected to chromatography and the active fraction was purified and found to be punicalagin with other inactive compounds such as ellagic acid and punicalin.[7] Punicalagin was found to inhibit the growth of Brazilian prevalent clone of six methicillin-resistant strains of *S. aureus.* Among all the fractions (hexane, chloroform, dichloromethane, ethyl acetate, butanol, water) of total ethanolic extract of fruit pericarp of pomegranate, ethyl acetate fraction containing a mixture of hydrolysable ellagitannins (α-lapachone, α-xyloidone, α-nor-hydroxylapachone) showed highest antibacterial activity against all strains of *S. aureus* including the methicillin-resistant strain.[64]

Methanolic extract of pomegranate was found to inhibit growth (0.01% v/v) of *S. aureus* FRI722. MIC for the same was at (1% v/v), and at a 0.05% (v/v) concentration of extract there was inhibition of Staphylococcal enterotoxin (SE) A production.[6] Olive extracts and phenolic compounds such as oleuropein have been shown to inhibit the growth and protein secretion of *S. aureus.*[66]

Punicalagin isolated from the fruit peel of pomegranate showed antimycobacterial activity against *M. tuberculosis* typus humanus ATCC 27294.[5] Ahmed and Beg[67] studied antimicrobial activity of ethanolic extracts of a few medicinal plants traditionally used in medicine against certain drug-resistant bacteria of clinical origin. Of these, pomegranate extract showed varied levels of antimicrobial activity against one or more test bacteria. Overall, broad-spectrum antimicrobial activity was observed in pomegranate. The presence of common phytocompounds in the plant extracts including phenols, tannins, and flavonoids as major active constituents may be responsible for these activities.

Pomegranate has also been used to control the bacterial diseases of plants. It has been reported that alcoholic extracts of fresh as well as dry skin of pomegranate are effective in inhibiting the growth of *Ralstonia solanacearum,* the causal agent of bacterial wilt of tomatoes.[68]

11.4 ANTIFUNGAL ACTIVITY

Punicalagin isolated from the fruit peel of pomegranate is reported to have antimicrobial activity against *Candida albicans.*[53] Fungistatic activity of pomegranate peel varied with test organisms[69] as it inhibited the growth of *Penicillium citrinum* for 8 days, *P. patulum* for 4 days, and *P. roquefortii* and *Aspergillus ochraceus* for 3 days. However, it had no effect on the growth of *A. flavus* and *A. parasiticus* (Table 11.1). Jia and Zia[70] have used pomegranate in fungicidal preparations. Jassim[71] reported antifungal and antiviral compositions comprising pomegranate extract. These compositions prevented the growth of fungus and virus but were not able to affect bacterial viability substantially. Ahmed and Beg[67] reported that a host of plant extracts including ethanolic extracts of pomegranate showed antifungal activity against *C. albicans.* The tough spongy portion of the pomegranate fruit, including fresh and sterilized juices, were reported to be effective against *Candida mycoderma.*[51]

In vitro studies have revealed that the extract of pomegranate inhibited the growth of oral bacteria and *Candida* species.[72,73] Recently, Vasconcelos et al.[73] reported that the gel prepared from the extract of the pericarp of fresh fruits of pomegranate is an antifungal agent against *Candida albicans.* It can be concluded that the extract of *P. granatum* may be used as a topical agent for the treatment of candidosis associated with denture stomatitis.

11.5 ANTIVIRAL ACTIVITY

In vitro screening has shown that some traditional medicinal herbs possess significant antiviral activities with no or limited adverse effects.[74,75] Many extracts from roots, stems, and fruits have been used in folk medicine for the treatment of viral diseases.[76,77] Extracts from some plant sources have in particular been shown to be effective against the influenza virus.[78,79] Many naturally occurring compounds, such as flavonoids, tannins, caffeic acid derivatives, terpenoids, and saponins, exhibit potent anti-HSV and anti-RSV activities *in vitro* or *in vivo.*[80–82] Among 21 medicinal herbs screened from southern mainland China, the aqueous extract of the fruit cortex of pomegranate was found to possess activity against all three human herpes simplex virus type 1 (HSV-1) (standard, clinical, and ACV-resistant) and human respiratory syncytial virus (RSV).[83] However, the IC_{50} was very low for the HSV-1 strains. The aqueous extract of pomegranate was also reported to possess anti-HSV-1 activity in other strains of HSV-1.[84] It is noteworthy that pomegranate contained abundant tannins.[85,86] Li et al.[83] have also demonstrated that the antiviral activity of this extract from pomegranate was possibly attributed to tannins, as the extract devoid of polyphenols (by passing through polyamide column) did not exhibit antiviral activity.

The hydroalcoholic extracts of whole fruits of pomegranate have exhibited high activity against the influenza virus.[87,88] Zhang et al.[89] evaluated antiviral activity of tannins from the pericarp of pomegranate against genital herpes virus (HVS-2). The tannins inhibited viral replication, demonstrated a killing effect, and blocked viral adsorption to target cells. Antiviral compositions comprising an extract of pomegranate rind were patented by Merck.[71] Stewart[90] reported that pomegranate rind water extract has a potent viricidal effect in combination with ferrous sulfate. The extracts from the hulls of pomegranate fruits have efficacy against viruses[91] including *Herpes simplex*,[89] poliovirus, and human immunodeficiency virus (HIV). The ethanolic extract from the leaf of pomegranate was found to have significant inhibitory effect on replication of Autographa California nuclear polyhedrosis virus.[92] Blocking virus entry is a primary target for microbicide development. Neurath et al.[93] isolated the HIV-1 entry inhibitors from pomegranate juice, which may bind strongly or irreversibly to the CD4 binding site on HIV-1 envelope gly-coprotein gp120, thereby preventing the entry of HIV-1 into the cells.

11.6 ANTIDIARRHEAL ACTIVITY

The hulls of pomegranate have been exploited in folk medicine throughout the world in the form of an aqueous decoction (i.e., boiling the hulls in water for 10 to 40 minutes), for dysentery and diarrhea, and also for stomatitis.[49,94] Kritikar and Basu[8] have reported the usefulness of pomegranate for the treatment of diarrhea. The studies of Das et al.[4] have established the efficacy of a pomegranate seed extract (methanol) as an antidiarrheal agent in mice. The pomegranate seed extract decreased intestinal propulsive movement in the charcoal-meal-treated model, and significantly inhibited PGE2-induced enteropooling. The above observation suggests that the pomegranate extract (in graded doses) reduced diarrhea by inhibiting gastrointestinal motility and PGE2-induced enteropooling. The inhibitory effect of the extract justifies its use in folk medicine and its use as nonspecific antidiarrheal agents. The underlying mechanisms for treatment of diarrhea appear to be spasmolytic and antienteropooling. Tannins are responsible for protein denaturation producing protein tannate, which reduces secretion from intestinal mucosa.[95] Pomegranate seed extract also contains tannin, which may produce antisecretory activity.

Segura et al.[96] evaluated the biological activity of alkaloids and tannins extracted from roots of pomegranate on axenic cultures from *Entamoeba histolytica* and *E. invadens*. Two milliliters of aqueous extract had higher activity on cultures from *E. histolytica* than *E. invadens* strains, producing growth inhibitions of about 100% and 40%, respectively. Alkaloid concentrations of 1 mg/ml had no amoebicide activity, but tannins at concentrations of 10 μg/ml for *E. histolytica* and 100 μg/ml for *E. invadens* were sufficient to produce a growth inhibition of about 100%. Tannic acid was also tested on the cultures of *E. histolytica* observing a high inhibitory activity on the growth; this effect, produced at 0.01 mg/ml, was similar to that observed with the tannins mixture. Dried pomegranate peels were also used for curing acute enteritis and dysentery.[97]

11.7 OTHER RELATED BIOACTIVITIES

The anthelmintic properties of various parts of the pomegranate plant, including the hulls of the fruit and the roots, have been documented.[98–100] Several important alkaloids including the pelletierrines have been characterized and shown to possess anthelmintic activity.[101] The most famous usage worldwide has been as a vermifugal or tanicidal agent[23,24] (i.e., a killer and expeller of intestinal worms). According to one account, the alkaoids contained in the root, tree bark, and to a lesser extent, fruit rinds, cause the "tapeworm to relax its grip on the wall of the intestine," thus allowing the weakened parasites to be easily expelled by a second herbal drug, one that is cathartic.[3] The fruit, bark, and stem of pomegranate contain a number of alkaloids belonging to the piperidine group, namely pelletierine, pseudopelletierine, iso-pelletierine, and methyl iso-pelletierine.[102,103] Hukkeri et al.[102] reported that the aqueous extract of the fruit rind of pomegranate was more active on tapeworms than on earthworms and roundworms. They reported that active principles are alkaloids, of which iso-pelletierine is the most potent tanicide. Schubert et al.[104] observed that pomegranate fermented juice and cold-pressed seed oil caused a significant inhibition of the eicosanoid pathway enzyme cyclooxygenase and lipooxygenase. The acetone (90%) extract of the pericarp of pomegranate caused a significant inhibition of carbonic dehydrogenase.[105] The extracts from the fruit hulls of pomegranate have shown antiparasitic activity against *Giardia*.[106]

It has been observed that pomegranate bark has potential molluscidal activity against harmful snails such as *Lymnaea acuminata,* the vector for *Fasciola hepatica* and *F. gigantica,* which are responsible for endemic fascioliasis in the cattle population.[107,108] The bark active fractions and the bark powder of pomegranate were found to inhibit the *in vivo* and *in vitro* activity of acetylcholinesterase, acid and alkaline phosphatase, $Na^+ K^+ATPase$, and lactic dehydrogenase in *Lymnaea acuminate*.[109] Thus the mortality in the snails by different preparations of pomegranate is contributed by several metabolic changes in the snail body.

11.8 STRUCTURE ACTIVITY RELATIONSHIPS

Several studies have shown the antimicrobial activities of the extracts from pomegranate may be indicative of the presence of some metabolic toxins or broad-spectrum antibiotic compounds. In the majority of the plants investigated, phenols and tannins were observed as the most common active constituents.[110] The major components in the extract of the fruits of pomegranate are tannins or polyphenols.[111] Nasr et al.[112] have reported that pomegranate peel contains ellagic acid, ellagitannins, and gallic acid. Pomegranate husk is rich in ellagitannins such as punicalagin and its isomers (2,3-hexahydroxydiphenoyl-4, 6-gallagylglucose), as well as lesser amounts of punicalin (4,6-gallagylglucose), gallic acid, ellagic acid, and ellagic acid glycosides (e.g., hexiside, pentoside, rhamnoside).[113–115] Ellagitannins and gallotannins have also been found in the heartwood of pomegranate.[59,86] Leaves and bark of this plant are known to be a rich source of tannins such as gallotannins and ellagitannins.[85,116–119] Ozkal and Dinc[120] reported the presence of tannins, anthocyanins, and flavonoids in pomegranate rind. Khaidarov et al.[121] extracted tannins from the rinds of pomegranate with the yield of 0.8 to 1.0%. Jayaprakasha et al.[122] suggest that the peels could be utilized for production of pharmaceutical preparations containing biologically active substances. Bark and root of the pomegranate tree also contain alkaloids, namely pelletierine, pseudopelletierine, N-methylpelletierine, seldridine, piperidine, isopelletierine, and norpseudopelletierine.[60,101,123] Plant alkaloids are known to possess antimicrobial activity[121,124] and alkaloids are known to have antimicrobial activity.[125]

Punicalagin is an ellagitannin containing a gallagyl moiety. It was originally isolated from the fruit peel of pomegranate and is reported to have antimicrobial activity.[5,7,53] Machado et al.[7] extracted pomegranate fruits with ethyl acetate and fractionated by chromatographic techniques to obtain ellagitannin punicalagin. The structure of the isolated compound was confirmed by ^{1}H NMR studies and the compound was found to be active against methicillin-resistant *Staphylococcus aureus* strains. Machado et al.[64] demonstrated that a mixture of hydrolysable ellagitannins (α-lapachone, α-xyloidone, α-nor-hydroxylapachone) isolated from the ethyl acetate fraction of the peels of pomegranate showed better antibacterial activity against all *S. aureus* strains tested including methicillin-resistant *Staphylococcus aureus* strains compared with the total ethanolic extract and other fractions, suggesting that this class of compounds (i.e., tannins) is responsible for the antibacterial activity observed in this plant. It is known that pomegranate is rich in tannins[85] and antimicrobial property of this substance is well established.[126] Polyphenols are known to form soluble complexes with proteins of high molecular weight. Thus, after being absorbed, the polyphenols will react with the protein moiety of cell enzymes (oxidoreductases) in the cytoplasm and in the cell wall. They may also bind to adhesions and so interfere with the availability of receptors on the cell surface.[127] In their study, Rani and Khullar[57] have also suggested that the antimicrobial activity of pomegranate was due to the presence of ellagitannins, gallotannins, and alkaloids. Prashanth et al.[21] have reported that among all the extracts, methanolic fraction from the pericarp of pomegranate was most effective against the test microorganisms and phytochemical screening gave positive tests for tannins and alkaloids for methanol extract. It appears that both tannins and alkaloids act synergistically against microorganisms. De et al.[52] reported that the possible reason for the antimicrobial properties of pomegranate seeds might be the presence of punicic acid. However, to establish it as a potent antimicrobial agent, the exact mode of action of the active principles must be studied in detail.

The major components in the extract of the fruits of pomegranate are tannins or polyphenols.[72,111] Vasconcelos et al.[73] reported that the gel prepared from the extract of the pericarp of fresh fruits of pomegranate is an antifungal agent against *C. albicans*. The specific mechanism of action of tannins against *Candida* is unclear, though it is suggested that tannins may act on the cell membrane because these compounds can precipitate proteins.[128] Tannins also inhibit many enzymes such as gycosyltransferases in *Streptococcus mutans* affecting the ability to attach to dental surfaces.[129] The adherence of *Candida* on acrylic surfaces is probably related to the presence of *S. mutans*.[128] Hence, polyphenols certainly interfere in salivary proteins and some oral bacterial enzymes. In addition, they may affect bacterial membranes and disturb bacterial coaggregation.

A wide variety of active phytochemicals, including the flavonoids, terpenoids, lignans, sulphides, polyphenolics, coumarins, saponins, furyl compounds, alkaloids, polyines, thiophenes, proteins, and peptides have been identified as antiviral agents.[125] The extracts from pomegranate showed antiviral activity against a good number of viruses, namely HSV-1, HSV-2, RSV, influenza virus, HIV, poliovirus, and polyhedrosis virus.[50,83,88–90,92] Li et al.[83] have demonstrated that the antiviral activity of the extract from pomegranate was possibly attributed to tannins, because the extract devoid of polyphenols (by passing through polyamide column) did not exhibit antiviral activity. Leaf ethanolic extract of pomegranate has a significant inhibitory effect on replication of *Autographa California* nuclear polyhedrosis virus by inhibition of virus-induced cytopathogenicity, synthesis of infected cell protein, and reduction in the virus titre.[92] The leaf tissue of pomegranate is rich in tannins, namely flavone glycosides and gallo- and ellagitannins.[85,116,117]

The HIV-1 entry inhibitors were found from pomegranate juice, which may bind strongly or irreversibly to the CD4 binding site on HIV-1 envelope glycoprotein gp120, preventing the virus entry.[93] Pomegranate juice contains several ingredients,[130] and those compounds have been isolated from other natural sources and were reported to possess anti-HIV properties, *viz.,* caffeic acid,[131] ursolic acid,[80] and catechin and quercetin.[132,133] However, Neurath et al.[93] observed that these compounds did not block HIV-1 envelope glycoprotein gp120-CD4 binding and did not adsorb to corn starch, unlike the entry inhibitor(s) from pomegranate juice. In fact, the supernatant, after treatment of pomegranate juice with starch and removal of the entry inhibitors, retained anti-HIV-1 activity and also inhibited infection by herpes virus type 1, unlike the HIV-1 entry inhibitors, which adsorbed onto starch. Thus, the antiviral activities in the supernatant appeared to be nonspecific and probably similar to those of extracts from pomegranate rind.[134,135]

11.9 CONCLUSIONS

The antimicrobial properties of plants have been investigated by a number of researchers worldwide. It has been observed that the extracts or purified compounds from the various parts of the pomegranate plant possess antibacterial, antifungal, antiviral, antidiarrheal, and other related activities. The active phytochemicals in pomegranate are found to be tannins and alkaloids. People have extensively applied the use of plants to heal diseases. Data from the literature reveal that the extracts or purified compounds from the various parts of the pomegranate plants have great potential for therapeutic treatment, even though they have not been completely investigated. Therefore, more studies need to be conducted to investigate compound-specific activities. Besides this, the search for new compounds is warranted for wide application of pomegranate in the treatment of infectious diseases caused by resistant microbes. The synergistic effect from the association of antibiotic with pomegranate extracts against resistant bacteria may lead to new choices for the treatment of infectious diseases. As the extracts were also effective against food-borne pathogens and spoilage bacteria, there are great possibilities for the use of pomegranate extracts as food biopreservatives.

ACKNOWLEDGMENT

We thank Dr. V. Prakash, director, and Dr. M.C. Varadaraj, head, Human Resource Development, CFTRI, Mysore, for their keen interest and constant encouragement.

REFERENCES

1. Vidal, A. et al., Studies on the toxicity of *Punica granatum* L. (Punicaceae) whole Fruit extracts, *J. Ethnopharmacol.*, 89, 295, 2003.
2. Madihassan, S., Outline of the beginnings of alchemy and its antecedents, *Am. J. Chinese Med.*, 12, 32, 1984.
3. Wren, R.C., *Potter's New Cyclopedia of Botanical Drugs and Preparations*, The C.W. Daniel Company, Essex, 1988, p. 90.
4. Das, A.K. et al., Studies on antidiarrheal activity of *Punica granatum* seed extract in rats, *J. Ethnopharmacol.*, 68, 205, 1999.
5. Asres, K. et al., Investigations on antimycobacterial activity of some Ethiopian medicinal plants, *Phytother. Res.*, 15, 323, 2001.
6. Gracious Ross, R., Selvasubramanian, S., and Jayasundar, S., Immunomodulatory activity of *Punica granatum* in rabbits 6 a preliminary study, *J. Ethnopharmacol.*, 78, 85, 2001.
7. Machado, T. et al., Antimicrobial ellagitannin of *Punica granatum* fruits, *J. Braz. Chem. Soc.*, 13, 606, 2002.
8. Kritikar, K.R. and Basu, B.D., *Indian Medicinal Plants*, 2nd ed., Bishen Singh and Mahendra Pal Singh, eds., Dehradun, Uttaranchal state, India, 1975, p.1084.
9. Nadkarni, A.K., *Nadkarni's Indian Material Medica*, vol. 1, Popular Prakashan, Bombay, 1976, 1031.
10. Yuyuan, C., Wound healing preparation, Chinese Patent CN 19939393104920, 1994.
11. Yusuph, M. and Mann, J., A triglyceride from *Punica granatum, Phytochemistry,* 44, 1391, 1997.
12. Stockwell, C., *Nature's pharmacy*, Century Hutchinson, London, Stewart, G., Antiviral and antifungal compositions comprise a mixture of ferrous salt and a plant extract of pomegranate rind, *Vibumum plicaum,* leaves or flowers, tea leaves or maple leaves in aqueous solution, Int. Search Rep. (patent pending), cited by Lee and Watson, 1998.
13. Thompson, W.A.R., ed., *Medicines from the Earth*, McGraw Hill, Maidenhead, United Kingdom, 1978, p. 49.
14. Robbers, J.M., Speedie, M., and Tyler, V., *Pharmacognosy and pharmacobiotechnology,* Williams and Wilkins, Baltimore, 1996, 1.
15. Jansen, A.M., Cheffer, J.J.C., and Svendsen, A.B., Antimicrobial activity of essential oils: a 1976–1986 literature review. Aspects of test methods, *Planta Med.*, 40, 395, 1987.
16. Saxena, G. et al., Antimicrobial constituents of *Rhus glabra, J. Ethnopharmacol.*, 42, 95, 1994.
17. Schubert, S.Y., Neeman, I., and Resnick, N., A novel mechanism for the inhibition of NF-κB activation in vascular endothelial cells by natural antioxidants, *FASEB J.*,16, 1931, 2002.
18. Argueta, A., Cano, L., and Rodarte, M., Atlas de las Plantas de la Medicina Tradicional, *FASEB Journal,* Mexicana, Mexico, D.F., Instituto Nacional Ingenista, Tomo, 1–3, 1994, 1786.

19. Baytelman, B., *Etnobotanica en el estado de Morelos. (Metodologia e introduccion al estudio de 50 plants de la zona norte del estado de Morelos),* Mexico, D.F., INAH, 1980, 280.

20. Gislene, G.F.N. et al., Antibacterial activity of plant extracts and phytochemicals on antibiotic resistant bacteria, *Brazilian J. Microbiology,* 31, 247, 2000.

21. Prashanth, D., Asha, M.K., and Amit, A., Antibacterial activity of *Punica granatum, Fitoterapia,* 72, 171, 2001.

22. Longtin, R., The pomegranate: nature's power fruit? *J. Nat. Cancer Institute,* 95, 346, 2003.

23. Zhicen, L., *Colour Atlas of Chinese Traditional Drugs,* Vol. 1, Science Press, Beijing, Peoples' Republic of China, 1987, 75.

24. Kapoor, L.D., *CRC Handbook of Ayurvedic Medicinal Plants,* CRC Press, Boca Raton, Florida, 1990, pp. 347–49.

25. Conner, D.E., *Antimicrobials in food,* Davidson, P.M. and Branen, A.L., eds., Marcel Dekker, New York, 1993, pp. 441–50.

26. Jayaprakasha, G.K. et al., Chemical composition of Turmeric oil — a byproduct from turmeric oleoresin industry and its inhibitory activity against different fungi, *Z. Naturforschung,* 56c, 40, 2001.

27. Jayaprakasha, G.K., Tamil Selvi, A., and Sakariah, K.K., Antibacterial and antioxidant activities of Grape *(Vitis vinifera)* seed extracts, *Food Res. Intern.,* 36, 117, 2003.

28. Jayaprakasha, G.K. et al., Antibacterial activity of *Citrus reticulata* peel extracts, *Z. Naturforschung,* 55c, 1030, 2000.

29. Negi, P.S. and Jayaprakasha, G. K., Antibacterial activity of grapefruit *(Citrus paradisi)* peel extracts. *Eur. Food Res. Tech.,* 213, 484, 2001.

30. Negi, P.S. and Jayaprakasha, G.K., Antioxidant and antibacterial activities of *Punica granatum* peel extracts, *J. Food Sci.,* 68, 1473, 2003.

31. Negi, P.S. and Jayaprakasha, G.K., Control of foodborne pathogenic and spoilage bacteria by garcinol and *Garcinia indica* extracts and their antioxidant activity, *J. Food Sci.,* 69, 61, 2004.

32. Ahmad, I., Mehmood, Z., and Mohammad, F., Screening of some Indian medicinal plants for their antimicrobials properties, *J. Ethnopharmacol.,* 62, 183, 1998.

33. Iyengar, M.A., *Study of crude drugs,* 2nd ed., College of Pharmaceutical Sciences, Manipal, 1985, 13.

34. Chopra, R.N. et al., *Glossary of Indian medicinal plants,* 3rd ed., Council of Scientific and Industrial Research, New Delhi, 1992, 7.

35. Harborne, J.B. and Baxter, H., eds., *Phytochemical dictionary, a handbook of bioactive compounds,* Taylor & Francis, 1995, p. 50.

36. Grosvenor, P.W., Supriono, A., and Gray, D.O., Medicinal plants from Riau Province, Sumatra, Indonesia. Part 2, Antibacterial and antifungal activity, *J. Ethnopharmacol.,* 45, 97, 1995.

37. David, M., Antimicrobial activity of garlic, *Antimicrobial agents chemother.,* 41, 2286, 1997.

38. Nimri, L.F., Meqdam, M.M., and Alkofahi, A., Antibacterial activity of Jordanian medicinal plants, *Pharm. Biol.,* 37, 196, 1999.

39. Ratnakar, P. and Murthy, P.S., Purification and mechanisms of action of antitubercular principle from garlic *(Allium sativum)* active against isoniazid susceptible and resistant *Mycobacterium tuberculae* H37RV, *Indian J. Clin. Biochem.,* 10, 14, 1995.

40. Saxena, V.K. and Sharma, R.N., Antimicrobial activity of essential oil of *Lantana aculeata. Fitoterapia,* 70, 59, 1999.

41. Saxena, K., Antimicrobial screening of selected medicinal plants from india, *J. Ethnopharmacol.*, 58, 75, 1997.
42. Emori, T.G. and Gaynes, R.P., An overview of nosocomial infections including the role of the microbiology laboratory, *Clin. Microbiol. Rev.*, 6, 428, 1993.
43. Pannuti, C.S. and Grinbaum, R.S., An overview of nosocomial infection control in Brazil, *Infect. Control Hosp. Epidemiol.*, 1, 170, 1995.
44. Boyce, J.M., Methicillin-resistant *Staphylococcus aureus* in hospitals and long-term care facilities: microbiology, epidemiology, and preventive measures, *Infect. Control Hosp. Epidemiol.*, 13, 25, 1992.
45. Santos, K.R.N. et al., DNA typing of methicillin-resistant *Staphylococcus aureus*: isolates and factors associated with nosocomial acquisition in two Brazilian university hospitals, *J. Med. Microbiol.*, 48, 17, 1999.
46. Cardoso, C.L., Teixeira, L.M., and Gontijo Filho, P.P., Antimicrobial susceptibilities and phage typing of hospital and non-hospital strains of methicillin-resistant *Staphylococcus aureus* isolated from hands, *Rev. Microbiol.*, 19, 385, 1988.
47. Aynechi, Y. et al., Screening of Iranian plants for antimicrobial activity, *Acta Pharmaceutica Suecia*, 19, 303, 1982.
48. Avirutnant, W. and Pogpan, A., The antimicrobial activity of some Thai flowers and plants, Mahido Univ., *J. Pharma. Sci.*, 10, 81, 1983.
49. Caceres, A. et al., Screening of antimicrobial activity of plants popularly used in Guatamela for the treatment of dermatomucosal diseases, *J. Ethnopharmacol.*, 20, 223, 1987.
50. Lee, J. and Watson, R.R., Pomegranate: a role in health promotion and AIDS? in *Nutrition, foods and AIDS*, Watson R.R., ed., CRC Press, Boca Raton, Florida, USA, 1998, 179.
51. Kirilenko, O.A. et al., Antibacterial properties of juice of various types of pomegranate, *Konservnaya I Ovoshchesushilnaya Promyshlennost*, 12, 12, 1978.
52. De, M., Krishna De, A., and Banerjee, A.B., Antimicrobial screening of some Indian spices, *Phytother. Res.*, 13, 616, 1999.
53. Burapadaja, S. and Bunchoo, A., Antimicrobial activity of tannins from *Terminalia citrine*, *Planta Medica*, 61, 365, 1995.
54. Navarro, V. et al., Antimicrobial evaluation of some plants used in Mexican traditional medicine for the treatment of infectious diseases, *J. Ethnopharmacol.*, 53, 143, 1996.
55. Perez, C. and Anesini, C., *In vitro* antibacterial activity of Argentinian folk medicinal plants against *Salmonella typhi*, *J. Ethnopharmacol.*, 44, 41, 1994.
56. Guivara, J.M., Chumpitaz, J., and Valencia, E., The *in vitro* action of plants on *Vibrio cholerae*, *Revista in gastroenterological del Peru*, 14, 27, 1994.
57. Rani, P. and Khullar, N., Antimicrobial evaluation of some medicinal plants for their potential against multidrug resistant *Salmonella typhi*, *Phytotherapy Res.*, 18, 670, 2004.
58. Jimeneg Misas, C.A. et al., Biological evaluation of Cuban plants IV, *Rev. Cubana Med. Trop.*, 31, 29, 1979.
59. El-Toumy, S.A.A., Marzouk, M.S., and Rauwald, H.W., Ellagi- and gallotannins *Punica granatum* heart wood, *Pharmazie*, 56, 823, 2001.
60. Ferrara, L. et al., Identification of the root of *Punica granatum* in galenic preparations using TLC, *Boll. Soc. Ital. Biol. Sper.*, 65, 385, 1989.
61. Voravuthikunchai, S. et al., Effective medicinal plants against enterohaemorrhagic *Escherichia coli* O157:H7, *J. Ethnopharmacol.*, 94, 49, 2004.
62. Alkofahi, A., Masaadeh, H., and Al-Khalil, S., Antimicrobial evaluation of some plant extracts of traditional medicine of Jordan, Alexandria, *J. Pharm. Sci.*, 28, 139, 1996.

63. Holetz, F.B., Pessini, G.L., and Sanches, N.R., Screening of some plants used in the Brazilian folk medicine for the treatment of infectious diseases, *Memorias do Instituto Oswaldo Cruz,* 97, 1027, 2002.

64. Machado, T.B. et al., *In vitro* activity of Brazilian medicinal plants, naturally occurring naphthoquinones and their analogues, against methicillin-resistant *Staphylococcus aureus, Int. J. Antimicro. Agents,* 21, 279, 2003.

65. Braga, L.C. et al., Pomegranate extract inhibits *Staphylococcus aureus* growth and subsequent enterotoxin production, *J. Ethnopharmacol.,* 96, 335, 2005.

66. Tranter, H.S., Tassou, S.C., and Nychas, G.J., The effect of the olive phenolics compound, oleuropein, on growth and enterotoxin B production by *Staphylococcus aureus, J. Applied Bacteriology,* 74, 253, 1993.

67. Ahmed, I. and Beg, A.Z., Antimicrobial and phytochemical studies on 45 Indian medicinal plants against multi-drug resistant human pathogens, *J. Ethnopharmacol.,* 74, 113, 2001.

68. Vudhivanich, S., Potential of some Thai herbal extracts for inhibiting growth of *Ralstonia solanacearum,* the causal agent of bacterial wilt of tomato, *Kamphaengsaen Acad. J.,* 1, 70, 2003.

69. Azzouz, M.A. and Bullerman, L.B., Comparative antimycotic effects of selected herbs, spices, plant components and commercial antifungal agents, *J. Food Prot.,* 45, 1298, 1982.

70. Jia, C. and Zia, C., A fungicide made from Chinese medicinal herb extract, Chinese Patent CN 1181187, 1998.

71. Jassim, S.A.A., Antiviral or antifungal composition comprising an extract of pomegranate rind or other plants and method of use, U.S. Patent 5840308, 1998.

72. Pereira, J.V., Atividade antimicrobianado estrato hidroalcoolico da *P. granatum* Linn, sobre microorganismos formadores da placa bacterium, Joao Pessoa, Federal University of Paraiba, Post-graduation in Dentistry. Thesis, 1998, 91.

73. Vasconcelos, L.C.S. et al., Use of *Punica granatum* as an antifungal agent against candidosis associated with denture stomatitis, *Mycoses,* 46, 192, 2003.

74. Fernandez-Romero, J.A. et al., *In vitro* antiviral activity of *Phyllanthus orbicularis* extracts against *Herpes simplex* virus type 1, *Phytother. Res.,* 17, 980, 2003.

75. Ma, S.C. et al., Antiviral Chinese medicinal herbs against respiratory syncytial virus, *J. Ethnopharmacol.,* 79, 203, 2002.

76. Sydiskis, R.J. et al., Inactivation of enveloped viruses by anthraquinones extracted from plants, *Antimicrobial agents and chemotherapy,* 35, 2463, 1991.

77. Taylor, R.S. et al., Anti-viral activities of Nepalese medicinal plants, *J. Ethnopharmacol.,* 52, 157, 1996.

78. Hayashi, K., Kamiya, M., and Hayashi, T., Virucidal effects of the steam distillate from *Houttuynia cordata* and its components on HSV-1, influenza virus and HIV, *Planta Medica,* 61, 237, 1995.

79. Nagai, T. et al., Mode of action of the anti-influenza virus activity of plant flavonoid, 5,7,4 trihydroxy 8-methoxyflavone from the roots of *Scutellaria baicalensis, Antiviral Res.,* 26, 11, 1995.

80. Ma, C. et al., Inhibitory effects of ursolic acid derivatives from *Cymorium songarcium,* and related triterpenes on human immunodeficiency viral protease, *Phytotherapy Res.,* 12, S138, 1998.

81. Kinjo, J. et al. Studies on antiherpetic drugs — Part 2. Anti-herpes virus activity of fabaceous triterpenoidal saponins, *Biol. Pharm. Bull.,* 23, 887, 2000.

82. Cheng, H.Y., Lin, C.C., and Lin, T.C., Antiherpes simplex virus type2 activity of casuarinin from the bark of *Terminalia arjuna, Antiviral Res.,* 55, 447, 2002.

83. Li, Y. et al., Antiviral activities of medicinal herbs traditionally used in southern mainland China, *Phytotherapy Res.,* 18, 718, 2004.

84. Namba, T., Shiraki, K., and Kurokawa, M., Development of antiviral agents from traditional medicines, in *Towards natural medicine research in the 21st Century,* Ageta, H., Aimi, N., Ebizuka, Y., Fujita, T., Honda, G., eds., Elsevier Science, Amsterdam, 1998, 67.

85. Hussein, S.A.M. et al., Tannins from the leaves of *Punica granatum, Phytochemistry,* 45, 819, 1997.

86. El-Toumy, S.A.A. and Rauwald, H.W., Two ellagitannins from *Punica granatum* heart wood, *Phytochemistry,* 63, 971, 2002.

87. Pena, B.R., BLBU: un extracto de frutos de *Punica granatum* L. con actividad contra el virus de la Influenza, Thesis for the Master in Microbiology Degree, Faculty of Biology, University of Havana, 1998.

88. Caballero, O. et al., Actividad inhibitory de extractos del fruto de *Punica granatum* sobre cepas del virus de la gripe, *Revista Cubana de Qu´Δmica XIII,* 106, 2001.

89. Zhang, J. et al., Anti-viral activity of tannin from the pericarp of *Punica granatum* L against genital *Herpes simplex* virus *in vitro, Chung Kuo Chung Yao Tsa Chin,* 20, 556, 1995.

90. Stewart, G., Antiviral and antifungal compositions comprise a mixture of ferrous salt and a plant extract of pomegranate rind, *Vibumum plicaum,* leaves or flowers, tea leaves or maple leaves in aqueous solution, *Int. Search Rep.* (patent pending, 1995), cited by Lee and Watson, 1998.

91. Mavlyanov, S. et al., Polyphenols of pomegranate peels show marked anti-tumor and anti-viral action, *Chem. Nat. Compounds,* 33, 98, 1997.

92. Erturk, O., Zihni, D., and Ali, O.B., Antiviral activity of some plant extracts on the replication of *Autographa californica* nuclear polyhedrosis virus, *Turk. J. Biol.,* 24, 833, 2000.

93. Neurath, A.R. et al., *Punica granatum* (pomegranate) juice provides an HIV-1 entry inhibitor and candidate topical microbicide, *BMC Infectious Diseases,* 4, 41, 2004.

94. Nagaraju, N. and Rao, K.N., A survey of plant crude drugs of Rayalaseema, Andhra Pradesh, India, *J. Ethnopharmacol.,* 29, 137, 1990.

95. Tripathi, K.D., *Essentials of Medical Pharmacology,* Jaypee Brothers Medical Publishers (P), New Delhi, 1994, 775.

96. Segura, J.J. et al., Growth inhibition of *Entamoeba histolytica* and *E. invadens* produced by pomegranate root *(Punica granatum L.), Archives de Investigaciones Medicas,* 21, 235 1990.

97. Shoutang, Z., Preparation and processing method for enterocleaning drug, Chinese Patent CN 19929292109615, 1994.

98. Caius, J.F. and Mhaskar, K.S., The correlation between the chemical composition of anthelmintics and their therapeutic value in connection with the hook worm inquiry in the Madras presidency. XIX. Drugs allied to them, *Indian J. Med. Res.,* 11, 353, 1923.

99. Prakash, V., Singhal, K.C., and Gupta, R.R., Anthelmintic activity of *Punica granatum* and *Artemisia siversiana, Indian J. Pharmacol.,* 12, 61A, 1980.

100. Naovi, S.A.H., Khan, M.S.Y., and Vohora, S.B., Antibacterial, antifungal and anthelmintic investigations on Indian medicinal plants, *Fitoterapia,* 62, 221, 1991.

101. Willaman, J.J. and Schubert, B.G., Alkaloid bearing plants and their contained alkaloids, ARS, USDA Tech. Bull. 1234, Supt. Documents, Govt. Print Off., Washington, DC, USA, 1961.

102. Hukkeri, V.I. et al., *In vitro* anthelmintic activity of aqueous extracts of fruit rind of *Punica granatum, Fitoterapia,* 64, 69, 1993.

103. *The wealth of India, raw materials,* Vol. VIII, Publication and Information Directorate, CSIR, New Delhi, India, 1993.
104. Schubert, S.Y., Lansky, E.P., and Neeman, I., Antioxidant and eicosanoid enzyme inhibition properties of pomegranate seed oil and fermented juice flavonoids, *J. Ethnopharmacol.,* 66, 11, 1999.
105. Ross, I.A., *Medicinal plants of the world: chemical constituents, traditional and modern medicinal uses,* Human Press, New Jersey, 1999, 273.
106. Ponce-Macotela, M., *In vitro* effect on *Giardia* of 14 plant extracts, *Revista de Investigacion Clinica,* 46, 343, 1994.
107. Tripathi, S.M. and Singh, D.K., Molluscicidal activity of *Punica granatum* bark and *Canna indica* Linn. root., *Braz. J. Med. Biol. Res.,* 33, 1351, 2000.
108. Tripathi, S.M. and Singh, D.K., Molluscicidal activity of *Punica granatum* and *Canna indica* combination with plant derived molluscides against harmful snail, *Malaysian Appl. Biol.,* 30, 25, 2001.
109. Tripathi, S.M. et al., Enzyme inhibition by the molluscicidal agent *Punica granatum* Linn. and *Canna indica* Linn. root, *Phyto therapy Research,* 18, 501, 2004.
110. Tanaka, T., Moriita, A., and Nanaka, G., Tannins and related compounds C III. Isolation and characterisation of new monomeric, dimeric and trimeric ellagitannins, calamanisanin and calamanins A, B, and C, from *Terminaliacaamansani.,* Chem. Pharm. Bull., 38, 60, 1991.
111. Haslam, E., Natural polyphenols (vegetable tannins) as drugs: possible modes of action, *J. Nat. Prod.,* 59, 205, 1996.
112. Nasr, C.B., Ayed, N., and Metche, M., Quantitative determination of the polyphenolic content of pomegranate peel, *Zeitschrz. fr Lebensmittel Unterschung und Forschung,* 203, 374, 1996.
113. Gil, M.I. et al., Antioxidant activity of pomegranate juice and its relationship with phenolic composition and processing, *J. Agric. Food Chem.,* 48, 4581, 2000.
114. Cerda, B. et al., Repeated oral administration of pomegranate ellagitannin punicalagin to rats for 37 days is not toxic, *J. Agric. Food Chem.,* 51, 3493, 2003.
115. Cerda, B. et al., Evaluation of the bioavailability and metabolism in the rat of punicalagin, an antioxidant polyphenol from pomegranate juice, *Eur. J. Nutr.,* 42, 18, 2003.
116. Nawwar, M.A.M., Hussein, S.A.M., and Merfort, I., NMR spectral analysis of polyphenols from *Punica granatum, Phytochemistry,* 36, 793, 1994.
117. Nawwar, M.A.M., Hussein, S.A.M., and Merfort, I., Leaf phenolics of *Punica granatum, Phytochemistry,* 37, 1175, 1994.
118. Tanaka, T., Nonaka, G., and Nishioka, I., Tannins and related compounds XI: revision of the structure of punicalin and punicalagin and isolation and characterization of 2-O-galloylpunicalin from the bark of *Punica granatum* L, Chem. Pharm. Bull., 34, 650, 1986.
119. Tanaka, T., Nonaka, G., and Nishioka, I., Tannins and related compounds XII: Isolation and characterization of novel ellagitannins, punicaroteins A, B, C and punigluconin from the bark of *Punica granatum* L, Chem. Pharm. Bull., 34, 656, 1986.
120. Ozkal, N. and Dinc, S., Evaluation of the pomegranate (*Punica granatum* L.) peels from the standpoint of pharmacy, *Ankara Univ Eczacilik Fak Derg,* 22, 21, 1994.
121. Khaidarov, K. et al., Tannins from pomegranate rind wastes, *Uzb. Khim. Zh.,* 6, 73, 1991.
122. Jayaprakasha, G.K., Singh, R.P., and Sakariah, K.K.A., Process for the isolation of antioxidants from pomegranate peels, Indian Patent Application No. 191/DEL/01/2001, 2001.

123. Neuhofer, H. et al., Alkaloids in the bark of *Punica granatum* L. (Pomegranate) from Yugoslavia, *Pharmazie,* 48, 389, 1993.
124. Gibbons, S., Anti-staphylococcal plant natural products, *Nat. Prot. Rep.,* 21, 263, 2004.
125. Jassim, S.A.A. and Naji, M.A., Novel antiviral agents: a medicinal plant perspective, *J. Applied Microbio.,* 95, 412, 2003.
126. Scalbert, A., Antimicrobial properties of tannins, *Phytochemistry,* 30, 3875, 1991.
127. Cowan, M.M., Plant products as antimicrobial agents, *Clin. Microbiol. Rev.,* 12, 564, 1999.
128. Branting, C., Sund, M.L., and Linder, L.E., The influence of *Streptococcus mutans* on adhesion of *Candida albicans* to acrylic surfaces *in vitro, Arch. Oral Biol.,* 34, 347, 1989.
129. Kakiuchi, N. et al., Studies on dental caries prevention by traditional medicines. VIII. Inhibitory effects of various tannins on glucan synthesis by glucosyltransferase, from *Streptococcus mutans, Chem. Pharm. Bull.,* 34, 720, 1986.
130. Poyrazoglu, E., Goekmen, V., and Artik, N., Organic acids and phenolic compounds in pomegranates (*Punica granatum* L.) grown in Turkey, *J. Food Composition and Analysis,* 15, 567, 2002.
131. Mahmood, N. et al., Inhibition of HIV infection by caffeoylquinic acid derivatives. *Antiviral Chem. Chemotherapy,* 4, 235, 1993.
132. Mahmood, N. et al., The anti-HIV activity and mechanisms of action of pure compounds isolated from *Rosa damascene, Biochem. BioPhys. Res. Commun.,* 229, 73, 1996.
133. De Tommasi, N. et al., Anti-HIV directed fraction of the extracts of *Margyricarpus setosus, Pharm. Biol.,* 36, 29, 1998.
134. http://www.aegis.com/news/re/1996/RE960310.html, Pomegranates could help in battle against AIDS, Reuters Ne-Media, Inc., March 10, 1996.
135. http://artsweb.bham.ac.uk/bmms/1996/03March96.html#Medical%20 breakthrough, Medical breakthrough, British Muslim Monthly Survey 1996, IV (36).

Section 3

Commercialization

12 Commercialization of Pomegranates: Fresh Fruit, Beverages, and Botanical Extracts

Navindra P. Seeram, Yanjun Zhang, and David Heber

CONTENTS

12.1 INTRODUCTION

The commercial potential and economic impact of pomegranates is enormous considering the different ways in which the fruit may be utilized. Consumption of pomegranates in the form of fresh fruit is very popular in various parts of the world, especially in Eastern and Middle Eastern countries where it is cultivated commercially in large quantities. Other commercial pomegranate products include jams, jellies, and wines. However, pomegranates are also popularly consumed as pomegranate juice (PJ). In the commercial juicing industry, large amounts of bioactive polyphenols contained in the fruit peel and membrane, consisting predominantly of punicalagin and its isomers, are extracted into PJ along with numerous other phytochemicals found in other parts of the fruit, such as its arils and seeds. Punicalagin

187

has been shown to be the major contributor to the potent antioxidant activities of PJ. It is noteworthy that punicalagin is present in commercial PJ whereas consumption of the edible fruit part, namely the arils, would not yield this potentially healthy phytochemical.

Although the economic impact of pomegranates on the food and beverage industry is huge, there is great potential for the use of pomegranate extracts as ingredients in functional foods, cosmeceuticals, nutraceuticals, phytoceuticals, and botanical dietary supplements (BDS). The consumption of dietary supplements consisting of botanicals or herbals, or vitamins and minerals, as adjuvants of complementary and alternative medicine (CAM) therapies, is extremely popular and used by more than 60% of Americans.[1,2] Over the past ten years, CAM therapies have grown into a multibillion-dollar industry in the U.S. alone, with more than 60 million Americans using herbal or other supplements and spending over $600 million annually on botanical products alone.[3] The widespread use of BDS is due to their being regarded as generally safe, arising out of their long history of traditional use; indeed, many botanical ingredients are included on the Generally Regarded as Safe (GRAS) list of the U.S. Food and Drug Administration (FDA).

This chapter therefore discusses the commercialization potential of pomegranate fruit, with a focus on its applications and impact on the BDS industry as a botanical extract.

12.2 POMEGRANATES AS FRESH FRUIT AND JUICE

The fresh fruit and juice market has grown steadily worldwide and has especially boomed in the U.S. over the past five years with increasing consumer awareness of the potential health benefits attributed to pomegranates. This is partially due to a growing body of both *in vitro* and *in vivo* studies, which demonstrate biological activities including antioxidant, antiatherogenic, anticancer, and anti-inflammatory properties of pomegranate phytochemicals. Currently, pomegranate fruits and commercially processed PJ are available from a wide variety of suppliers in the world. The following list (arranged alphabetically by suppliers) shows some of the PJ suppliers in the U.S.:

1. Naturally Pomegranate™ Liquid Dietary Supplement, from AgroLabs (http://www.naturallypomegranate.com)
2. Wonderful Pomegranate Juice Power, by Brownwood Acres (http://www.brownwoodacres.com)
3. Organic Pomegranate Juice, by California Academy of Health, Inc. (http://www.caoh.org/pomegranate.html)
4. Natural Pomegranate Juice Concentrate, by Dynamic Health Labs (http://www.dynamic-health.com)
5. Pomegranate Juice Concentrate, by Jarrow Formulas (http://www.jarrow.com)
6. Pomegranate Juice, by Knudsen & Son (http://www.grainsandgreens.com)

7. Pomegranate Juice Concentrate, by Lakewood (http://www.lakewood-juices.com)
8. Fresh Pomegranate Juice, by Pomegranate Farm (http://www.pomegranate-farm.com)
9. Pomegranate Juice, by Pom Wonderful[R] (http://www.pomwonderful.com)
10. Pomegranate Simple Syrup, by Sonoma Syrup Co. (http://www.sonomasyrup.com)

12.3 POMEGRANATES AS BOTANICAL EXTRACTS

As previously discussed, there is great potential for the use of pomegranate extracts as botanical ingredients in dietary supplements, cosmeceuticals, and so forth. As in the case of the fresh-fruit and PJ industry, there are also a large number of botanical ingredient suppliers of pomegranate extracts. The following list (arranged alphabetically by suppliers) shows some of the pomegranate botanical ingredient suppliers in the U.S.:

1. PomEllagic™, by Cyvex (http://www.cyvex.com)
2. Cardio-Edge 120 Veggie Caps, by Douglas Labs (http://www.douglaslabs.com)
3. Living Multi (Optimal Men's Formula), 180 Caplets, by Garden of Life (http://www.gardenoflifeusa.com)
4. PomElla™, by Geni Herbs (http://www.geniherbs.com)
5. Pomegranate Green Tea — 50 Tea Bags, by Herbal Remedies (http://www.herbalremedies.com)
6. Life Extension Mix Powder, by iHerb (http://www.iherb.com)
7. Pomegranate Extract Capsules (30), by Lifes Vigor (http://www.lifesvigor.com)
8. Pomegranate Power Caps, 90 cap (seeds oil), by Mother Nature (http://www.mothernature.com)
9. Pomegranate Extract 30 caps, by Nature's Herbs (http://www.naturesherbs.com)
10. Standard Pomegranate Fruit Extract, by Naturex (http://www.naturex.com)
11. Pomegranate V capsules (60), by Nature's Way (http://www.naturesway.com)
12. CardioGranate 60 Vegetarian Capsules and EstraGranate 60 Vegetarian Capsules, by Pomegranate Health (http://www.pomegranatehealth.com)
13. PomEllagic Plus, by Pure Encapsulations (http://www.purecaps.com)
14. Pomegranate Extract Caps and PhytoEstrogen & Ipriflavones Capsule, by Solaray (http://www.affordablesolaray.com)
15. OxyPhyte® Pomegranate Extract, by RFI Ingredients (http://www.rfiingredients.com)
16. Pomegranate Extract Tablets and Phyto-Estrogen Cream, by Source Naturals (http://www.sourcenaturals.com)
17. PomAgic™, by Triarco (http://www.triarco.com)

12.3.1 Issues with the Dietary Supplement Industry

There are several issues with the dietary supplement industry. Under the 1994 Dietary Supplement Health and Education Act (DSHEA), a manufacturer is responsible for ensuring that its dietary supplement is safe before it is marketed. Generally, manufacturers do not need to register with the U.S. Food and Drug Administration (FDA) or get approval before selling dietary supplements, The FDA can take action against potentially fallacious claims or unsafe dietary supplements only after a product reaches the market. Therefore, it is the manufacturer's responsibility to ensure that product label information is accurate and not misleading. Unfortunately, complex mixtures of plant extracts that are used as active ingredients in BDS are not standardized, are commonly adulterated, and vary immensely depending on geological origin, environmental factors, harvesting practices, and so on. The lack of commercial or relevant reference standards, suitable analytical methods, and contamination with pesticides and other closely related but sometimes unknown and harmful plant materials further complicate matters. Therefore, a combination of well-designed basic and clinical studies is needed to evaluate the quality, safety, bioavailability, and herb–drug interactions of a BDS. All of the above factors should be considered and carefully investigated for the development of pomegranates and other botanicals as dietary supplements.

12.3.2 Traditional Medicinal Uses of Pomegranates

Various parts of the pomegranate such as its bark, leaves, flowers, and fruit have been used as folk medicines for centuries by many cultures, documented as far back as 1550 B.C.[4] For example, pomegranates have been used for the treatment of diabetes, conjunctivitis, hematuria, diarrhea, anthelmintic, dysentery, and so forth.[5–7] There are several commercial herbal formulations utilizing pomegranate extracts in various parts of the world such as Brazil, India, and China.[8,11] In India, almost all parts of the pomegranate plant are used for the treatment of various ailments. The flower is used for diarrhea and dysentery[9,10] and the peels have been used to prepare wound-healing material. Pomegranate seeds that contain steroids such as estrone and coumestrol have also been used for a variety of chemotherapeutic purposes.[12] Because various parts of the pomegranate tree (bark, stems, leaves, and roots) and fruits (arils, rind, pith, pericarp, and seeds) have been utilized by ancient cultures for health-care purposes, their modern-day medical use is also commonly practiced. Currently, most of the studies attributing health benefits to pomegranates have been conducted on the fruit (including its arils, peel, and seeds) in beverage forms (juice and wines) and as extracts. By far, the most abundant phytochemicals present in the pomegranate fruit (arils and peels) are polyphenols, while the seeds contain fatty acids and triglycerides. Studies focusing on these dietary pomegranate phytochemicals are therefore necessary.

12.4 CHEMICAL MARKERS OF POMEGRANATES

Phytochemical marker compounds serve as authentic reference standards that can be used in an established, reliable, rigorous, and validated analytical method for the quality control (QC) of botanical extracts and food products. Common analytical instruments used in the QC testing of botanical extracts include high-performance liquid chromatography (HPLC) with ultraviolet (UV) and mass spectral (MS) detectors, and gas chromatography (GC) with flame ionization and MS detectors. The presence and levels of a suitable chemical marker compound, usually within an established and acceptable range, have a variety of applications as follows: (1) evaluation for authenticity, (2) standardization, (3) consistency (i.e., batch-to-batch variations in lot numbers), and (4) shelf-life and stability studies.

There are several issues involved in establishing suitable chemical markers for botanical extracts. The "ideal" chemical marker is (1) well-characterized and identified using physical properties (such as melting point), chemical and spectral (such as nuclear magnetic resonance [NMR] or HPLC) methods, (2) easily available, inexpensive, and accessible, and (3) unique to a particular botanical, which is not easily achievable given the ubiquity of secondary metabolites in plants. It should also be noted that a suitable chemical marker might not be the biologically active ingredient(s) and therefore may have no relevance to the anecdoctal, purported, or reported bioactivities of a botanical extract. Botanical extracts are distinguished from single pharmaceutical drugs due to additive or synergistic effects of multiple and sometimes related compounds present in their particular matrix.

Suitable chemical markers for pomegranate fruit extracts are divided into hydrophilic (water-soluble) and lipophilic (fat-soluble) phytochemicals, as discussed below.

12.4.1 Hydrophilic Markers of Pomegranates

Hydrolysable tannins (HTs) consist of ellagitannins (ETs) and gallotannins (GTs) and are the predominant polyphenols found in pomegranate fruit, accounting for 92% of its antioxidant activity.[13] These compounds occur in the fruit peel (pericarp, membrane, and pith) and are extracted into pomegranate juice (PJ) during commercial hydrostatic processing methods. As previously discussed, the major pomegranate HT is an ET known as punicalagin (or punicalagins, since it exists as isomers).[13,14] The punicalagins can reach levels as high as 2 g/L of PJ, depending on processing methods and fruit cultivar, and are responsible for about half of the total antioxidant capacity of PJ.[13] ETs and GTs are susceptible to enzymatic and nonenzymatic hydrolysis releasing ellagic acid (EA) or gallic acid molecules, respectively.[15] The ability of ETs to release EA on hydrolysis is commonly used to detect and estimate ET presence and content in the absence of an authentic ET reference standard.[16] ETs are found in a number of fruits and particularly in berries such as strawberries, raspberries, black raspberries, and blackberries.[16,17] Therefore, EA is not unique to pomegranates and is rather a general marker of the presence of ETs. In conclusion, given the abundance of punicalagins in pomegranate peels,[18,19] we propose that punicalagin is a better chemical marker compound for pomegranate extracts than EA. Although other plants have been reported to contain punicalagins, such as leaves of *Cistus salvifolius*[20] and fruits of *Terminalia chebula*,[21] among commonly consumed edible fruits, only pomegranates contain punicalagins. Other candidates that could serve as suitable "unique" chemical markers for pomegranate instead of EA are related ETs such as punicalin and gallagic acid.[18] Our laboratory and others have also shown that the punicalagins are largely responsible for the antioxidant activity of pomegranates.[3,13] We have also shown that punicalagin shows pro-apoptotic, COX-2, and NFκ-B inhibitory activities, while EA alone is not active.[22] It should also be noted that although pomegranate fruits also contain several flavonoids such as anthocyanins (cyanidin, pelargonidin, and delphinidin glycosides) and flavonols (quercetin, kaempferol, and luteolin), these would be less-preferred chemical markers for phytochemical analyses of pomegranate extracts due to their ubiquity in fruits and vegetables.

12.4.2 LIPOPHILIC MARKERS

Lipophilic pomegranate phytochemicals are found in the seeds and consist of fatty acid and triglycerides of which punicic acid, a c9, t11, c13-conjugated linolenic acid, reaches levels of >60% of the seed oil. We propose that punicic acid is the best-suited candidate to serve as a chemical marker of pomegranate seed oil.[24]

12.5 STUDIES OF SEVEN COMMERCIAL POMEGRANATE BOTANICAL EXTRACTS

Our laboratory obtained seven commercial pomegranate extracts labeled supplements A to G in order to obscure manufacturer's identities. The extracts were obtained as free-flowing powders and were analyzed prior to their expiration dates. Each lot number was separately sampled and analyzed three times. Using authentic punicalagin standard isolated in our laboratory as previously reported,[18] and ellagic acid (purchased from Sigma, USA), the following studies were conducted.

12.5.1 PHYTOCHEMICAL ANALYSES

Using previously reported HPLC methods,[18] extracts A to G were analyzed for punicalagins and ellagic acid contents (Table 12.1). Of the seven extracts surveyed, three contained punicalagins (B: 42%, D: 10.9%, and E: 26.5%), two contained trace amounts of punicalagins (F and G), and no punicalagins were detected in two of the extracts (A and C). All of the extracts contained EA, ranging from 4 to 90%, and three extracts contained high percentages of EA (C: 86.5%, D: 72.7%, G: 90.0%).

In conclusion, our survey of seven pomegranate botanical extracts showed that only three extracts contained punicalagins, the proposed chemical marker for pomegranates. The high levels of EA coupled with the absence of punicalagins in some of the extracts may be due to hydrolytic conversion of punicalagins and other related ETs to EA during extraction or processing methods. In addition, spiking of botanical extracts with commercial or synthetic EA, as well as EA obtained from other plant sources, may be a tempting practice that could account for the high levels of observed EA.

12.5.2 STABILITY STUDIES

Because extract B contained the highest amount of punicalagins and its HPLC profile showed the most similarities to PJ,[18,23] it was selected for stability studies as follows.

TABLE 12.1
Punicalagins and Ellagic Acid Content in Seven
Commercial Pomegranate Botanical Extracts

Pomegranate Extracts	Fruit Part	Punicalagins (%)	Ellagic Acid (%)
A	Skins	nd	47.8
B	Whole fruit	42.0	4.8
C	Whole fruit	nd	86.5
D	Skins	10.9	72.7
E	N/A	26.5	40.2
F	N/A	Trace	47.0
G	N/A	Trace	90.0

HPLC analyses were conducted as previously reported.[18]

N/A = not available; nd = not detected

Extract B was aliquoted into transparent vacuum-sealed plastic bags and analyzed by HPLC for shelf-life stability at days 1, 7, 15, 30, 60, 90, 120, 150, and 180. At all time points, the punicalagin content remained constant showing its stability at these shelf-life conditions of room temperature and light exposure.

The extract was also subjected to extreme conditions of varying pH (3, 7, and 10) and thermal conditions (ambient, boiling by reflux, and microwave). The results are summarized in Table 12.2.

At ambient temperature, the punicalagin content was constant at pH 3 and 7. However, when the temperature was increased to boiling, the punicalagin in the extract at pH 7 started decomposing into an unidentified compound corresponding to a shoulder in its HPLC profile. As previously discussed, because ET are HTs, it is possible that punicalagin can be converted to the related ETs, punicalin and gallagic acid.[18] At a basic pH 10, the punicalagins a and b peaks (Figure 12.1a), corresponding to the α and β anomers of punicalagin respectively, were inverted in concentration. The EA content also increased, confirming that ETs produce EA as a breakdown product (Figure 1b).

At boiling (reflux) temperatures, punicalagin at pH 3 was stable while the punicalagin at pH 7 again started decomposing into an unidentified peak, possibly punicalin. At the basic pH 10, punicalagin further decomposed into a compound identified by LC-MS as gallagic acid.[18] The EA content also increased.

Under microwave conditions, the punicalagins were stable at pH 3 and 7 but started decomposing to gallagic acid and EA at pH 10 (Figure 1b).

In conclusion, punicalagins are stable at ambient, reflux, and microwave conditions at low pH 3. However, under neutral and basic conditions, coupled with extreme temperatures, punicalagin decomposes to form gallagic acid and EA as its major breakdown products.

TABLE 12.2
Extreme Conditions Stability Studies Conducted on a Pomegranate Botanical Extract Exposed to Varying pH and Temperatures

pH	Conditions	Qualitative Results
3	Ambient (180 min)	No degradation
7	Ambient (180 min)	No degradation
10	Ambient (180 min)	Punicalagin a and b peaks have inverted relative concentrations; EA peak area increase ~2X
3	Reflux (20 min)	No degradation
7	Reflux (20 min)	New shoulder peak observed (unidentified)
10	Reflux (20 min)	Punicalagin a and b disappear; new large peak identified as gallagic acid (m/z: 601)[18]; EA peak area increases ~2.75X
3	Microwave (5 min)	No degradation
7	Microwave (5 min)	No degradation
10	Microwave (5 min)	Punicalagin a and b peaks disappear; New large peak identified as gallagic acid (m/z: 601)[18]; EA area increases ~5X

The decomposition products were identified by HPLC and LC-MS/MS as previously reported.[18]

p = punicalagins; EA = ellagic acid; nd = not detected.

12.6 CONCLUSIONS

Currently the commercialization of pomegranates as fresh fruit and beverages has a well-established market. However, given the abundance of bioactive punicalagins in the peel of the pomegranate fruit, there remains a viable industry that can utilize the spent by-product material generated from the commercial juice industry to generate ET-rich extracts that have great potential for use as botanical ingredients.[18] Therefore the utilization of pomegranate extracts as ingredients may have immense commercial application in the functional food, beverage, cosmeceutical, nutraceutical, and phytoceutical industries. However, as discussed above, there remain inherent issues that are encountered in the development of botanical extracts with precisely defined characteristics and consistent quality and that can be used as ingredients. Nevertheless, the advantages of having well-characterized and standardized pomegranate botanical extracts are enormous. For example, these extracts can be used in animal and human studies where administration of sufficient quantities of a particular bioactive ingredient, such as punicalagin, is required to achieve desired biological endpoints. Furthermore, studies can be designed using a pomegranate botanical extract that contains multiple compounds with pharmacodynamic properties that are different from those of a pharmaceutical drug or isolated single-chemical compounds. Our laboratory and others continue to demonstrate the additive or synergistic effects of multiple phytochemicals found in botanical extract matrices.[25,26] On a cautionary note, it should be reemphasized that the use of BDS and herbal medicines

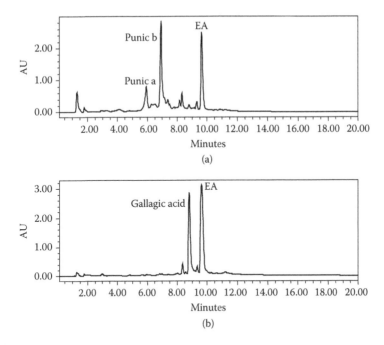

FIGURE 12.1 HPLC chromatograms of pomegranate botanical extract before (a) and after (b) extreme conditions stability studies. The extract contains the major pomegranate ellagitannin, punicalagins (punicalagin a and b), and ellagic acid (EA). The major degradation compounds are EA and gallagic acid, identified by LC-MS/MS.

should be considered in association with an individual's diet, medical history, and prescribed treatments, especially drug therapy, since herb–drug interactions may be dangerous in some cases.

REFERENCES

1. Kessler, R.C. et al., Long term trends in use of complementary and alternative medical therapies in the United States, *Ann. Intern. Med.,* 135, 262, 2001.
2. Brevoort, P., The booming US botanical market: a new overview, *Herbalgram,* 44, 33, 1998.
3. Seeram, N.P. et al., Dietary polyphenols derived from pomegranates are potent antioxidants: evaluation in various *in vitro* models of antioxidation. In *228th National Meeting of the American Chemical Society; Agricultural and Food Division, Lipid oxidation and antioxidants: chemistry, methodologies and health effect,* 2004, Philadelphia.
4. Wren, R.C., *Potter's new cyclopedia of botanical drugs and preparations,* C.W. Daniel, Essex, 1988, 90.
5. Das, A.K. et al., Studies on antidiarrheal activity of *Punica granatum* seed extract in rats, *J. Ethnopharmacol.,* 68, 205, 1999.
6. Asres, K. et al., Investigations on antimycobacterial activity of some Ethiopian medicinal plants, *Phytother. Res.,* 15, 323, 2001.

7. Gracious, R.R., Selvasubramanian, S., and Jayasundar, S., Immunomodulatory activity of *Punica granatum* in rabbits ó a preliminary study, *J. Ethnopharmacol.*, 78, 85, 2001.

8. Machado, T. et al., Antimicrobial ellagitannin of *Punica granatum* fruits, *J. Braz. Chem. Soc.*, 13, 606, 2002.

9. Kritikar, K.R. and Basu, B.D., *Indian Medicinal Plants*, 2nd ed., Bishen Singh and Mahendra Pal Singh, eds., Dehradun, Uttaranchal state, India, 1975, 1084.

10. Nadkarni, A.K., *Nadkarni's Indian Material Medica*, vol. 1, Popular Prakashan, Bombay, 1976, 1031.

11. Yuyuan C., Wound healing preparation, Chinese Patent CN19939393104920, 1994.

12. Yusuph, M. and Mann, J., A triglyceride from *Punica granatum. Phytochemistry*, 44, 1391, 1997.

13. Gil, M.I. et al., Antioxidant activity of pomegranate juice and its relationship with phenolic composition and processing, *J. Agric. Food Chem.*, 48, 4581, 2000.

14. Doig, A.J. et al., Isolation and structure elucidation of punicalagin, a toxic hydrolyzable tannin, from *Terminalia oblongata, J. Chem. Soc., Perkin Trans. 1*, 8, 2317, 1990.

15. Haslam, E., *Plant polyphenols: Vegetable tannins revisited*, Cambridge University Press, Cambridge, 1989, 230.

16. Amakura, Y. et al., High performance liquid chromatographic determination with photodiode array detection of ellagic acid in fresh and processed fruits, *J. Chromatogr. A*, 896, 87, 2000.

17. Seeram, N.P. et al., Identification of phenolics in strawberries by liquid chromatography electrospray ionization mass spectroscopy, *Food Chem.*, 2005 (in press).

18. Seeram, N.P. et al., Rapid large-scale purification of ellagitannins from pomegranate husk, a by-product of the commercial juice industry, *Sep. Purif. Technol.*, 41, 49, 2005.

19. Seeram, N.P. and Heber, D., Purification and uses of pomegranate ellagitannins, 2004, Patent Application: USPTO No. 60/556,322.

20. Saracini, E. et al., Simultaneous LC-DAD and LC-MS determination of ellagitannins, flavonoid glycosides, and acyl-glycosyl flavonoids in *Cistus salvifolius* L. leaves, *Chromatographia*, 62, 245, 2005.

21. Juang, L.-J. and Sheu, S.-J., Chemical identification of the sources of commercial fructus chebulae, *Phytochem. Anal.*, 16, 246, 2005.

22. Adams, L. et al., Pomegranate juice, total pomegranate tannins and punicalagin suppress inflammatory cell signaling in colon cancer cells, *J. Agric. Food Chem.*, 2005, in press.

23. Seeram, N.P., Lee. R., and Heber, D., Bioavailability of ellagic acid in human plasma after consumption of ellagitannins from pomegranate (*Punica granatum* L.) juice, *Clin. Chim. Acta*, 348, 63, 2004.

24. Kohno, H. et al., Pomegranate seed oil rich in conjugated linolenic acid suppresses chemically induced colon carcinogenesis in rats, *Cancer Sci.*, 95, 481, 2004.

25. Seeram, N.P. et al., Total cranberry extract versus its phytochemical constituents: antiproliferative and synergistic effects against human tumor cell lines, *J. Agric. Food Chem.*, 52, 2512, 2004.

26. Seeram, N.P. et al., *In vitro* antiproliferative, apoptotic and antioxidant activities of punicalagin, ellagic acid and a total pomegranate tannin extract are enhanced in combination with other polyphenols as found in pomegranate juice, *J. Nutr. Biochem.*, 16, 360, 2005.

Section 4

Plant Growth and Improvement

13 Pomegranates: A Botanical Perspective

David W. Still

CONTENTS

13.1 ANCIENT HISTORY

In many cultures the pomegranate figures prominently in various myths concerning life and various human aspirations. In Greek mythology, it served as a symbol of the indissolubility of marriage; in Persian mythology, Isfandiyar ate a pomegranate and became invincible; in Judaism, the number of pomegranate seeds in a single fruit are said to number 613, one for each of the Bible's 613 commandments; in Buddhism, the pomegranate is one of the three blessed fruits; in Chinese ceramic art, the pomegranate is associated with fertility, abundance, posterity, numerous and virtuous offspring, and a blessed future; in Christianity and Bedouin tribes, it is associated with fertility; in Islam, the Koran describes a heavenly paradise that contains pomegranates.[1]

Depictions of pomegranate fruits have been found adorning different architectural and artistic designs throughout much of history. Scholars have long debated whether the pomegranate or date palm was the tree of life portrayed in various archaeological artifacts from Mesopotamia, the Levant, and India. A recent analysis from a botanical archaeologist concluded the tree of creation or immortality that recurs in many archaeological materials from the first through third millennium before present (BP) is actually the Egyptian locust.[2]

Pomegranate designs have been used in artistic works throughout ancient history. For example, paper wrappers, which were designed primarily to protect or adorn unbound pamphlets before they were bound together with other pamphlets in a more permanent larger volume, are known to have been produced during the fifteenth and early sixteenth centuries in only three towns, Augsburg, Ferrara, and Venice. One of the oldest surviving wrappers, dating from 1486, takes the form of a pomegranate as part of its pattern.[3] The pomegranate was well established by the late Bronze Age and it is found, perhaps as artwork, or for its symbolism, in a number of Egyptian Eighteenth Dynasty tomb paintings and funerary garlands.

Where written records are lacking, contents from shipwrecks can provide a wealth of information about the cultural history of early civilizations. Seafaring in the ancient Mediterranean began more than 10,000 years ago and evidence of the dependence on ships to transport merchandise is reflected in texts dating from the end of the fourth millennium BP. A ship carrying the finest luxury goods of the late Bronze Age sank off the coast of Ulu Burun, Turkey, in the late fourteenth century BP. Its recent discovery has yielded great insight into cultural life during Egypt's Eighteenth Dynasty and the LH IIIB period in Greece. The ship contained ceramic containers in which more than 1000 archaeobotanical samples were retrieved. These contained almonds, acorns, pine nuts, olives, fig and grape seeds, and pomegranates, indicating the importance of the fruit during this time.[4] Archaeobotanical evidence from military sites shows that pomegranates were brought to Central Europe by Roman soldiers during their occupation of this area.[5] Because of climatic conditions, pomegranates could not be grown in these areas and as a consequence remained an imported luxury item.[5]

In addition to the symbolic nature of pomegranates, its chemical properties made it useful for a number of widely disparate purposes, ranging from perfumes to birth control. A classical list of perfume ingredients from Pliny's *Natural History* included pomegranate rind, while its juice was used as an astringent to prepare oil to receive a scent. Pomegranates were highly valued fruits that were associated with symbols of fertility and rebirth in the religious iconography of the Late Bronze Age.[6] Ironically, however, pomegranate peels or rinds were included in recipes for oral contraceptives but were more commonly used as a vaginal suppository abortifacient during classical and medieval times.[7] In Greek mythology, the pomegranate's effects on fertility was noted, and in both ancient and modern Indian literature its use as an abortifacient has been reported.[7] In his review of plant-derived oral contraceptives and abortifacients, Riddle[7] cited several laboratory experiments that confirmed the pomegranate's effects on fertility, with the primary evidence being that rats and guinea pigs that were fed pomegranates had fewer pregnancies than those not fed pomegranates. An early (1933) analysis of pomegranate leaves indicated high amounts of estrogens, while estrone was later identified from pomegranate seeds using thin-layer chromatography.[8] More recently, the estrogenic compounds luteolin, quercetin, and kaempferol were detected in pomegranate rind using APCI MS/MS.[9] On the basis of the reports, workers concluded that infertility was probably caused by the relatively high levels of estrogens. However, a recent analysis of pomegranate seeds, whole fruits, and commercial juice preparations using HPLC-PDA or GC-MS detection methods failed to find any steroidal hormones (namely, estrone, estradiol, or testosterone).[10] This latest analysis suggests that either the earlier conclusions of pomegranates containing steroids were wrong or steroidal hormone levels vary among cultivars. Alternatively, steroid content of plants may change seasonally. If pomegranates do not have high levels of steroidal hormones, the cause of lowered fertility in animal models remains unknown.

13.2 DOMESTICATION

Pomegranates are one of the earliest fruit species to be domesticated. In general, the process of domestication begins simply by removing plants from their native habitat and growing them in an area spatially separated from the naturally occurring population so as to preclude gene flow. If the seeds are not replenished by collecting from plants throughout the species' geographic range, genetic drift will occur. Over time, the harvested collection will no longer contain the full genetic repertoire of the naturally occurring population. The process of domestication occurred over thousands of years, and in many cases the amount of genetic changes that have occurred by the action of early man far surpasses the efforts of plant breeders over the last 100 years.[11]

The morphological changes that occur during domestication are often so great that a domesticated species may look little like its nondomesticated progenitor. One of the most thoroughly studied examples of domestication has been maize (*Zea mays* L.), which was domesticated from teosinte in Mesoamerica over a period of 4000 to 5000 years. The vegetative and reproductive morphology of maize little resembles the native teosinte that is still found in the Mexican highlands. In pomegranate, however, such stark differences between the domesticated and nondomesticated forms do not occur. The suspected progenitor of pomegranate is very similar in appearance to the domesticated form, differing mainly in the size of the fruit.

During the process of domestication, selection pressures imparted by early man often resulted in plants with larger seeds and fruit, nondehiscent fruits and seeds, and changes in seed or fruit color.[11,12] The earliest plants to be domesticated were the cereals, which were domesticated 10,000 to 12,000 BP in the Fertile Crescent, an area of land extending from the Nile Valley along the Mediterranean to northern Syria, down the Tigris and Euphrates valleys toward the Persian Gulf. With the exception of figs and grapes, fruit trees were only domesticated within the last 5000 years or so.[12] Pomegranates are thought to have been domesticated in the Transcaucasia-Caspian region,[12] specifically in northeastern Turkey and the south Caspian regions,[13] although other opinions exist as to the area of origin. The spread of pomegranates can be estimated by archaeological evidence, as described in Spiegel-Roy.[13] Carbonized pips and fragments of pomegranate peels have been found from early Bronze Age Jericho and Arad, and remains of *Punica* species have been found in Nimrud. By the Middle Ages, pomegranate grew throughout the Levant and appeared in Egypt during the Middle Kingdom. Many wild groves can still be found thriving in the bottom of gorges throughout the Mediterranean and Central Asian region.[14] Because of pomegranate's wide geographic distribution and its long history of use, a very wide range of diversity exists.[14]

It is interesting to note that many plants were domesticated for their symbolic value or for use in religious ceremonies rather than for food. For example, Native Americans used maize, pumpkins, and gourds in their religious ceremonies. Selections were made on the basis of color, as each color had symbolic meaning. Although rind color in pomegranates varies from yellow to deep red, and seed color from cream to deep red, the deep red color is most widely favored in such disparate cultures as the U.S. (California) and the Middle East.[15]

Why some plants are domesticated and others are not is an interesting question, but the biology and genetics of the plant often yields clues. The ease of domestication is greatly facilitated by an annual growth habit, meaning the seed-to-flowering process occurs within one year. While most early domesticated species were annual cereals that reproduce by self-fertilization, fruit trees are perennials and predominantly outcrossing. Thus, we may suspect fruit trees were domesticated later than annuals largely because of their much longer reproductive cycle, in which typically 5 years elapse before an appreciable amount of fruit is produced. Pomegranates appear to outcross at approximately 13%,[16] so seedlings are not "true to type." As a result, pomegranates exhibit great plant-to-plant variation. Early domestication of pomegranate was likely facilitated by vegetative reproduction of individual plants, presumably selected because of their desirable characteristics.[13]

13.3 BOTANICAL DESCRIPTION

Pomegranates (*Punica granatum* L.) have been variously placed in the Lythraceae family, or Punicacea, depending on the taxonomist and whether they are considering morphological or molecular data. The former is a large family comprised of 31 genera and 620 species, whereas Punicacea contains only two species, *Punica granatum* and *Punica protopunica* Balf. f., with the latter endemic to the small island of Socotra off the coast of Yemen. When defined in the broadest sense, the Lythraceae family comprises four subfamilies, with one, Punicoideae, sometimes regarded as a satellite family of Lythraceae *sensu stricto*. The generalized morphology of the family, together with the very distinctive genera, suggests that the family is of great age.[17]

Pomegranate grows as a shrub or small tree up to 5 m; its leaves are simple and opposite, with entire margins and a deciduous habit. The flowers are bisexual, with radial symmetry, and the fruit is classified as a berry that is divided into irregular sections, each containing many edible seeds, with the seeds surrounded by a fleshy pulp, called an aril.

13.4 CULTURAL CONDITIONS

Pomegranate is cultivated throughout the world in subtropical and tropical areas in many different microclimatic zones. Approximately 100,000 ha are cultivated worldwide, with a total annual production of 800,000 metric tons.[14] Pomegranates may be propagated by seeds, which readily germinate, or vegetatively in the spring by hardwood cuttings, and in the summer by softwood cuttings. Pomegranates are generally thought to be drought resistant because once established they can survive in semiarid regions with no additional irrigation. Like most fruit trees, however, pomegranates are very sensitive to even slight water deficits, which negatively affect stem, root, and leaf growth and formation.[18] Repeated or prolonged water deficits will greatly reduce the amount of fruit produced. From a commercial standpoint, pomegranates should not be considered a drought-tolerant plant and irrigation should be scheduled so as not to stress the plant. With respect to water deficits, the most sensitive phase of a plant's growth cycle occurs during pollination and fertilization; it is critically important not to incur water deficits during these phases. Further, water deficits will likely result in splitting of the fruit, making it unsuitable for the fresh market. Although virtually all pomegranate production occurs in arid and semiarid regions, pomegranates are also grown in humid areas such as Florida and Hawaii. Relatively less water is lost by evapotranspiration in areas of high humidity because of lower water vapor deficits between the leaf and atmosphere, but the fruit is more prone to disease and is of lower quality than that produced in arid regions.[19] Systematic large-scale screening of pomegranate germplasm for fruit production under water deficits has not been reported, nor has screening against disease resistance, but it is likely that differences exist among genotypes.

13.5 GENETIC DIVERSITY

More than 500 pomegranate varieties are known around the world but only 50 are commonly grown.[14] The trend of utilizing only a relatively small portion of the total genetic diversity for commercial purposes is not unique to pomegranates. The very processes of domestication and breeding of elite cultivars resulted in genetic bottlenecks in most crops surveyed.[20] That is, the amount of genetic diversity of our modern hybrids, cultivars, and varieties is a fraction of what previously existed, or now exists in the wild progenitors. The sustainability of agriculture depends on broadening the genetic base, especially to produce crops in marginally arable areas.[21] Fortunately, breeders have recognized this danger and are actively seeking to maintain genetic diversity by conserving germplasm.

Genetic diversity is defined and limited by the diversity that exists in gene pools. A gene pool exists in the collective population of plants that occur throughout the range of the species, termed *in situ* germplasm, or it may be conserved in specialized collections where it is known as *ex situ* germplasm conservation. The primary gene pool is composed of plants of the same or closely related species, while secondary gene pools include more distantly related species that often display partial F_1 sterility; a tertiary gene pool includes more distantly related species that upon crossing will produce severe F_1 sterility.[22] With genetic engineering, and the theoretic ability to transfer any gene into any unrelated organism, a quaternary gene pool can be defined. Because pomegranate is an underutilized minor fruit tree crop, little information exists that would define compatible gene pools; thus the primary gene pool remains the most accessible gene pool for improving pomegranates.

Several international efforts have been initiated to collect, preserve, and evaluate pomegranate germplasm.[14] One of the richest areas for secondary gene pools for pomegranate exists in Turkmenistan, which possesses rich agricultural biodiversity, but because of severe and widespread ecological degradation the wild pomegranates are in danger of imminent extirpation.[23] The largest *ex situ* pomegranate germplasm

TABLE 13.1
Worldwide *Ex Situ* Germplasm Collections of Pomegranate
(*Punica granatum* L.)

Country	Center	Location	Number of Accessions	Reference
Albania	Research Institute of Fruit Trees and Vineyard	Tirana	5	33
Cyprus	ARI, Plant Genetics Resources and Herbarium	Nicosia	3	33
EMFTS[a]	11 locations	Italy, Spain	116	24
France	CIRAD-FLHOR	Capesterre Belle-Eau	2	33
Germany	Institute of Crop Science, Federal Res. Center	Braunschweig	2	33
	Inst. For Production Nutr. of World Crops	Witzenhausen	ng[b]	33
Hungary	University of Hort. Food Industry	Budapest	3	33
Israel	Newe Ya'ar Research Center	Haifa	29	33
Portugal	National Fruit Breeding Station	Alcobaca	5	33
Russia	Vavilov Research Institute of Plant Industry	St. Petersburg	800	33
Tunisia	Gabes	Tunisia	>60	34
Turkey	Aegean Agricultural Research Inst.	Izmir	158	33
	Aegean University	Izmir	13	33
Turkmenistan	Garrygala Research Station	Garrygala	1117	14
Ukraine	Nikita Botanical Garden	Yalta	370	35
U.S.	Fruit and Nut Germplasm Center	Davis, CA	59	36
Uzbekistan	Schroeder Uzbek Research Center	Tashkent	ng[b]	37

[a] European Minor Fruit Tree Species, EC Project GENRES 29.
[b] Number of accessions not given.

in the world is held at the Garrygala Research Station in Turkmenistan, with 1117 accessions (Table 13.1). Scientists from Syria, Libya, Tunisia, and Turkmenistan are working to collect and characterize this germplasm. Other international efforts, supported by the European Commission, include collection and evaluation efforts by Italy, Greece, Spain, and France.[24] A modest germplasm collection is also held in the U.S. (Table 13.1).

In order to prevent duplication and accurately assess the variation that exists in these collections, each accession will need to be characterized not only in terms of its morphological variation, but should also include a genome-wide survey of the genetic diversity. Because of the expense associated with this, and the fact that this is a minor fruit crop, funding for such a venture is limited. Genetic assessment of

TABLE 13.2
Range of Physiochemical Fruit Properties of Pomegranate Germplasm

	Mars	Al-Kahtani	Barone et al.	Manohar et al.
Fresh weight (g)	196–674	99–428	255–393	78–155
Fruit length (mm)	47–96	—	68–81	50–61
Fruit diameter (mm)	58–111	—	83–102	50–70
External color	6–11[a]	3.4–8.8[e]	—	—
Sepal number	5.8–7.6	—	—	—
Calyx length (mm)	12.4–21.9	—	—	—
Calyx diameter (mm)	18.5–33.1	—	—	—
Juice volume (cm³/100 g seeds)	72.3–100.3	—	—	—
Juice color	2–11.2[b]	—	—	—
pH	3.1–4.6	3.3–4.3	3.9–4.2	—
Total soluble solids	133.5–169[c]	14.1–16.3[f]	12.9–16.9[f]	11.0–15.0[f]
Acidity	0.25–2.1[d]	0.2–1.4[f]	0.34–1.95[f]	0.76–2.30[f]
Skin (rind) weight (g)	76–282.1	—	–	25–75
Skin (rind) thickness (mm)	2.8–6.1	—	4.2–23.5	3.0–4.5
Arils/fruit	—	—	359–600	232–501
100-aril weight (g)	—	—	27.2–60.5	15.2–22.8
Shelf life	—	35–112[g]	—	—

[a] 14-point scale; greenish yellow (01) to dark purplish red (14)
[b] 16-point scale; cream (01) to dark red (16)
[c] g/l
[d] g/l expressed as malic acid content
[e] Subjective, nonlinear sensory 10-point scale; yellow to dark red
[f] %
[g] Storage, in days, at 20°C, 47% RH
Source: Data from Mars and Marrakchi,[32] Al-Kahtani,[15] Barone et al.,[38] and Manohar et al.[39]

unsequenced plants is largely accomplished by use of anonymous molecular markers, especially amplified fragment-length polymorphism markers. Thus, genetic diversity (or similarity) can only be implied from these methods and valuable genes that potentially contribute to increased drought resistance, different phytochemical profiles, or disease resistance, may not be identified by use of anonymous markers. More sophisticated approaches using mapping populations and marker-assisted breeding would facilitate the identification of valuable loci and genes and the development of new varieties.

Of the more than 2500 accessions that exist (Table 13.1), only a very small fraction has been evaluated and none of these have included genetic fingerprinting or direct assessments of genetic diversity; however, in the accessions evaluated to date, large ranges were observed in the physical and chemical attributes of the fruit (Table 13.2). Although morphological characters are greatly affected by the environment, these results suggest significant diversity exists for which, presumably, there is an underlying genetic basis. For example, Al-Kahtani[15] evaluated the postharvest quality of 11 pomegranate cultivars commonly grown in the Middle East.

He noted significant differences among cultivars in fruit weight, percent edible portion of fruit, percent juice, sensory evaluation (including color, flavor, mouth feel, and overall acceptability), compressive strength and strain as a measure of freshness during storage, and chemical characteristics (including pH, total soluble solids, acidity, vitamin C, and total sugar; Table 13.2). In some cases the differences between the best- and worst-performing varieties in a given trait were more than 700%. Other collections have been examined, and examples of the range in physiochemical traits among the accessions are given in Table 13.2. Within a collection of 35 sweet varieties grown in Spain (sweet varieties were defined as those having a total acidity between 0.16 and 0.42 g/100 g fresh fruit), each accession displayed unique organic acid profiles, and as many as sixfold differences within specific organic acids were observed.[25] Total acidity, however, differed by only twofold among the cultivars. The same trends were observed with lipid content and fatty acid composition of seed oil from clones of seven pomegranate accessions grown in commercial orchards in Spain. Each cultivar had unique lipid and fatty acid profiles, and specific lipids or fatty acids varied widely among cultivars.[26]

Breeding objectives for pomegranate have included easily measured phenotypic traits such as hardness of seed, rind thickness, seed content, and fruit color,[27] but in general, any trait can be improved that exhibits genetic variation with high to moderate heritability. As data become available concerning associations between specific phytochemicals (e.g., anthocyanins, lipid profiles) and health benefits, it is likely that elite cultivars can be developed that are enriched for that phytochemical. A common method among breeders when creating new varieties is to estimate heritability and identify transgressive segregants from a cross, meaning an individual plant that outperforms either parent. This is also known as heterosis or hybrid vigor. Karale and Desai[27] investigated the amount of heterosis that existed among F_1 progeny from crosses among 17 parents (four serving as males, thirteen as females). In the 24 traits examined, significant transgressive segregation was identified in 19 traits. Significant heterosis was not observed in fruit length, diameter, aril percent, rind percent, or yield per plant. These results do not preclude the possibility that the aforementioned traits cannot be improved if all the available germplasm were screened. It is interesting to note that the traits displaying little heterosis were the same traits selected for during domestication of pomegranate.[13]

Most of the traits in pomegranates we are interested in improving are controlled by multiple genes, and significant genetic interaction with the environment will likely occur. Comparisons among pomegranate accessions can only be made when all plants are grown on the same plot and experience the same environmental and cultural conditions. Similarly, comparisons should be made when the fruits are at a comparable developmental stage as the phytochemical content changes as the fruits mature.[28] Fruits from trees grown in different geographical areas in Israel differed in their phytochemical content, perhaps because of climatic effects,[28] while other researchers noted that total anthocyanin content of juice differed by maturation stage and geographic location.[29] These traits could potentially be improved by making crosses with different genetic material within the primary gene pool, or by crossing with closely related species from a secondary gene pool. The results of such wide crosses are difficult to predict. Examples exist in which horticultural traits have been improved following wide crosses and the genetic basis for the improvement has been elucidated. For example, the wild progenitor of the domesticated tomato *(Lycopersicon esculentum)* likely had fruit less than one cm in diameter and only a few grams in weight.[30] All wild *Lycopersicon* species examined have small-fruit alleles while all domesticated varieties have large-fruit alleles. The genes responsible for the increased fruit size, which had its origin in the small-fruited progenitors, was identified first as a quantitative trait locus and later mapped to a single gene, ORFX, that controls the carpel cell number and is similar to a human oncogene.[31] This gene, which is a major contributor to fruit size, is essentially a domestication gene. It is likely that genetic improvement through wide crosses can be made in pomegranates, and as more pomegranate accessions are screened and mapping populations become available, genes contributing to valuable horticultural traits will be identified.

Perhaps as a result of its long history of use and the movement of germplasm throughout the Mediterranean, pomegranate germplasm appears to lack any geographic patterns of variation. Mars and Marrakchi[32] examined 14 fruit morphometric and chemical attributes (acidity and total soluble solids) of 30 pomegranate accessions collected from different geographic locations in Tunisia. Using principal component analysis, they showed that fruit weight, length, diameter, and skin weight accounted for most of the variation among cultivars. Although significant differences existed for each trait among the accessions, these differences could not be attributed to the accession's geographical origin; because only a small sampling from the geographic range was made, it would be premature to conclude that no geographically related genetic or morphological variation exists in pomegranates.

13.6 SUMMARY

Pomegranates have a long and exceptionally colorful history, having been embraced by a number of different cultures, while at the same time remaining a minor horticultural tree crop. Yet, despite its wide geographic distribution across several continents, very little information is available concerning its genetic origin, centers of diversity, or evaluation of *in situ* and *ex situ* genetic diversity, and relatively little breeding has been performed. On the other hand, because it has not been subjected to strong selection pressures during domestication and very little breeding has occurred that would lead to genetic bottlenecks, much apparent genetic diversity still remains. Thus, given enough economic incentive, which is likely to be driven by potential phytochemical benefits of the fruit, these obstacles can be overcome and the potential to develop elite pomegranate varieties exists.

REFERENCES

1. Langley, P., Why a pomegranate, *BMJ*, 321, 1153, 2000.
2. McDonald, J.A., Botanical determination of the Middle Eastern tree of life, *Econ. Bot.*, 56, 113, 2002.
3. Hirsch, R., The decoration of a 1486 book wrapper and its reappearance in 1531, *Stud. Renaissance*, 6, 167, 1959.

4. Haldane, C., Direct evidence for organic cargoes in the late Bronze Age, *World Arch.,* 24, 248, 1993.

5. Bakels, C. and Jacomet, S., Access to luxury foods in Central Europe during the Roman period: the archaeobotanical evidence, *World Archaeology,* 34, 542, 2003.

6. Ward, C., Pomegranates in eastern Mediterranean contexts during the Late Bronze Age, *World Arch.,* 34, 529, 2003.

7. Riddle, J.M., Oral contraceptives and early-term abortifacients during classical antiquity and the Middle Ages, *Past Present,* 132, 3, 1991.

8. Heftmann, E., Ko, S.-T., and Bennett, R.D., Identification of estrone in pomegranate seeds, *Phytochem.,* 231, 1337, 1966.

9. van Elswijk et al., Rapid dereplication of estrogenic compounds in pomegranate *(Punica granatum)* using on-line biochemical detection coupled to mass spectrometry, *Phytochem.,* 65, 233, 2004.

10. Choi, D.W. et al., Identification of steroid hormones in pomegranate *(Punica granatum)* using HPLC and GC-mass spectrometry, *Food Chem.,* in press, 2005.

11. Hancock, J.F., *Plant evolution and the origin of crop species,* 2nd ed., CABI Publishing, Cambridge, MA, 2004, chapter 7.

12. Harlan, J.R., *Crops and man,* 2nd ed., Am. Soc. Agron. Inc., Madison, WI, 1992, chapters 2, 3.

13. Zohary, D. and Spiegel-Roy, P., Beginnings of fruit growing in the old world, *Science,* 187, 319, 1975.

14. IPGRI, Regional report CWANA 1999–2000, International Plant Genetic Resources Institute, Rome, Italy, 2001.

15. Al-Kahtani, H.A., Intercultivar differences in quality and postharvest life of pomegranates influenced by partial drying, *J. Amer. Soc. Hort. Sci.,* 117, 100, 1992.

16. Jalikop, S.H. and Kumar, P.S., Use of a gene marker to study the mode of pollination in pomegranate *(Punica granatum* L.), *J. Hort. Sci.,* 65, 221, 1990.

17. Graham, S.A., Crisci, J.V., and Hoch, P.C., Cladistic analysis of the Lythraceae *sensu lato* based on morphological characters, *Bot. J. Lin. Soc.,* 113, 1, 2000.

18. Badizadegan, M., Growth of pomegranate *(Punica granatum* L.) as affected by soil moisture tension, *J. Hort Sci.,* 50, 227, 1975.

19. LaRue, J.H., Growing pomegranates in California, DANR Pub. Leaflet 2459, Reprinted July 1980, Univ. CA.

20. Tanksley, S.D. and McCouch, S.R., Seed banks and molecular maps: unlocking genetic potential from the wild, *Science,* 277, 1063, 1997.

21. Jaradat, A.A., in *Transformations of Middle Eastern natural environments: legacies and lessons,* Coppock, J., and Miller, J.A., eds., Bulletin Series Number 103, Yale School of Forestry and Environmental Studies, Yale University, New Haven, Connecticut, 1998.

22. Michelmore, R.W., The impact zone: genomics and breeding for durable disease resistance, *Cur. Op. Plant Biol.,* 6, 397, 2003.

23. Anonymous, Biodiversity assessment for Turkmenistan, Task order under the biodiversity and sustainable forestry IQC (BIOFOR), submitted by Chemonics International Inc., submitted to USAID Central Asian Republics Mission, Almaty, Kazakhstan, 2001.

24. Bellini, E. and Giordani, E., Conservation, evaluation, exploitation and collection of minor fruit tree species, EC Project GENRES 29, 1998, last accessed via internet on 4 April 2006, http://www.unifi.it/project/ueresgen29/

25. Melgarejo, P., Salazar, D.M., and Artes, F., Organic acids and sugars composition of harvested pomegranate fruits, *Eur. Food Res. Technol.,* 211, 185, 2000.

26. Melgerejo, P. and Artes, F., Total lipid content and fatty acid composition of oilseed from lesser known sweet pomegranate clones, *J. Sci. Food Ag.,* 80, 1452, 2000.

27. Karale, A.R. and Desai, U.T., Study of heterosis for fruit characters in inter-cultivar crosses of pomegranate (*Punica granatum* L.), *Indian J. Genet.,* 60, 191, 2000.

28. Shulman, Y., Fainberstein, L., and Lavee, S., Pomegranate fruit development and maturation, *J. Hort. Sci.,* 59, 265, 1984.

29. Gil et al., Influence of cultivar, maturity stage and geographical location on the juice pigmentation of Tunisian pomegranates, *Z. Lebensm. Unters. Forsch.,* 201, 361, 1995.

30. Rick, C.M., Zobel, R.W., and Fobes, J.F., Four peroxidase loci in red-fruited tomato species: genetics and geographic distribution, *Proc. Nat. Acad. Sci. USA,* 71, 835, 1974.

31. Frary, A. et al., *fw2.2:* A quantitative trait locus key to the evolution of tomato fruit size, *Science,* 289, 85, 2000.

32. Mars, M. and Marrakchi, M., Diversity of pomegranate (*Punica granatum* L.) germplasm in Tunisia, *Gen. Res. Crop Evol.,* 46, 461, 1999.

33. Frison, E.A. and Servinsky, J., *ECP/GR directory of European Institutions Holding Crop Genetic Resources Collections, Vol. 1, Holdings,* Int. Plant Genetic Resources Institute, 1995.

34. Mars, M. and Marrakchi, M., Conservation and characterization of the genetic resources of pomegranate (*Punica granatum* L.) in Tunisia, *Plant Gen. Resources Newslet.,* 114, 35, 1998.

35. Yezhov et al., Genetic resources of temperate and subtropical fruit and nut species at the Nikita Botanical Gardens, *Hort. Science,* 40, 5, 2005.

36. USDA/ARS National Germplasm Repository for Fruit and Nut Crops. Germplasm list may be accessed via the following link, last accessed 6 September 2005: http://www.ars-grin.gov/ars/PacWest/Davis/punica.html.

37. Mirzaev, M.M. et al., The Schroeder Institute in Uzbekistan: Breeding and germplasm collections, *Hort. Science,* 39, 917, 2004.

38. Barone, E. et al., Preliminary observations on some Sicilian pomegranate (*Punica granatum* L.) varieties, in *Production, processing and marketing of pomegranate in the Mediterranean region: advances in research and technology,* Melgarejo-Moreno, P., Martínez-Nicolás, J.J., and Martínez-Tomé, J., eds., Zaragoza, CIHEAM-IAMZ, 2000, 137.

39. Manohar, M.S., Tikka, S.B.S., and Lal, N., Phenotypic variation and its heritable components in some biometric characters in pomegranate (*Punica granatum* L.), *Ind. J. Hort.,* 38, 187, 1981.

14 Postharvest Biology and Technology of Pomegranates

Adel A. Kader

CONTENTS

14.1 INTRODUCTION

The pomegranate, *Punica granatum* L., belongs to the *Punicaceae* family. It is one of the oldest known edible fruits and is associated with ancient civilizations of the Middle East. From its origin in the area now occupied by Iran and Afghanistan, the pomegranate spread east to India and China and west to Mediterranean countries (Turkey, Egypt, Tunisia, Morocco, and Spain). Spanish missionaries brought the pomegranate to the Americas in the 1500s.[1,2] The primary commercial pomegranate-growing regions of the world are the Near East, India and surrounding countries, and southern Europe. Nearly all production in the U.S. is centered in the southern San Joaquin Valley of California,[2] with about 4000 hectares (10,000 acres) of pomegranates (predominately of the "Wonderful" variety). The fruit is consumed fresh or processed into juice, syrup, jams, or wine.[2-6] This chapter provides an overview of postharvest biology and technology of pomegranates in relation to maintaining their quality between harvest and fresh consumption or processing.

14.2 MORPHOLOGICAL CHARACTERISTICS

The pomegranate fruit is nearly round, varying in diameter from about 6.25 to 12.5 cm (2.5 to 5 inches) with a prominent and persistent calyx and a hard, leathery skin (rind or husk). Skin color varies from yellow overlaid with light or dark pink to bright red, depending on variety.[3-6] Botanically, the pomegranate is classified as a berry, but the edible portion develops not from the seedbox wall but from the outer seedcoat. The edible part is the pulp (aril) surrounding the seeds. The pulp, together with the locular septa (membranous walls and white, spongy tissues), represent the whole pericarp. The arils are filled with juicy red, pink, or whitish (depending on variety) pulp. In each aril (juice sac) there is one angular, soft or hard seed. Although the fruit's leathery skin appears to provide protection from physical damage and water loss, this is not the case. The skin can easily be scuffed from abrasions and has many microcracks and other openings that facilitate water loss. Thus, pome-granates should be handled with as much care as apples during harvesting and postharvest handling.

14.3 COMPOSITION AND COMPOSITIONAL CHANGES
DURING MATURATION AND RIPENING

"Wonderful," the most widely grown pomegranate variety in California, has a deep purple-red skin color with a glossy appearance. The arils and juice are a deep crimson color with good flavor (due to high contents of sugars and acids);[3] seeds are small and tender, and the rind is of medium thickness.[2,7] In general, varieties that have whitish or pinkish arils (such as the "Mollar" grown in Spain) are usually sweeter than those with purplish or dark crimson arils because the latter varieties contain higher concentrations of organic acids.[8] The edible portion (arils) of pomegranate is about 55–60% of total fruit weight and consists of about 75–85% juice and 15–25% seeds.[9,10]

Citric acid is the predominant organic acid, and titratable acidity ranges from 1 to 2% (fresh weight basis).[7,10] Glucose and fructose are the main sugars, which range from 14 to 17% (fresh weight basis).[7,10] The common anthocyanins in pome-granate juice are the 3-glucosides and 3,5-glucosides of delphinidin, cyanidin, and pelargonidin.[11-15] Other phenolic compounds in pomegranate include ellagic acid derivatives and hydrolysable tannins (punicalagin, punicalin).[16] There is a strong positive correlation between total phenolics and antioxidant activity of pomegranate.[16,17] However, above a certain concentration, phenolic compounds can render the juice less desirable because of astringency.

Titratable acidity decreases and soluble solids (mainly sugars) content, pH, and red color intensity of the juice increase with pomegranate fruit maturation and ripening.[7,10,18] For example, California-grown "Wonderful" pomegranates picked in mid-October had an average soluble solids content of 18.1% and a titratable acidity of 1.58%, whereas those harvested in late September averaged 17 and 1.8% respectively.[7] There was no consistent relationship between the extent of red coloration of the skin and red color intensity of the arils. Differences in soluble solids, juice color, percent edible portion, and percent extractable juice were small among fruits of various sizes. Large fruits (more than 250 grams) were generally lower in titratable acidity than smaller fruits.[7]

14.4 MATURITY INDICES

The pomegranate fruit reaches full maturity (ripeness) within 4.5 to 6 months after bloom, depending on climactic conditions.[2,10,19] The fruits should be harvested before they become overripe and crack (split) open, especially under rainy conditions. Maturity indices are variety dependent and include external skin color (changes from yellow to red) and juice color, acidity, and soluble solids content.[7,9,18,19] The maximum titratable acidity may be 1% in sweet varieties and 1.5 to 2% in sweet–sour varieties. Minimum soluble solids vary from 15 to 17%. The minimum maturity indices for California-grown "Wonderful" pomegranates are red juice color equal to or darker than Munsel color chart 5R-5/12 and titratable acidity below 1.85%.[7,18,20]

14.5 QUALITY INDICES

Pomegranate fruit quality depends on the following indices:

- Freedom from internal and external decay.
- Freedom from preharvest defects (such as cracking/splitting and sunburn,[21] which cause dark brown to black discoloration of the affected skin area) and defects that may occur during harvesting and handling (such as surface abrasions, cuts, and impact bruising).
- Skin color and smoothness.
- Aril color intensity and uniformity.
- Fruit size — may be considered a quality index, depending on the intended use of the pomegranates.
- Flavor depends on sugar/acid ratio, which varies among cultivars. Soluble solids contents above 17% and total phenolics contents below 0.25% are desirable for optimal levels of sweetness and astringency, respectively.[7,20]

14.6 POSTHARVEST PHYSIOLOGY

14.6.1 RESPIRATION AND ETHYLENE PRODUCTION RATES

Pomegranate fruits have a relatively low respiration rate that declines with time during storage after harvest.[18,7] The ranges of respiration (carbon dioxide production) rates for California-grown "Wonderful" pomegranates were 2–4, 4–8, and 8–18 ml/kg hr at 5°C, 10°C, and 20°C, respectively, while ethylene production rates remained below 0.2 microliter per kilogram per hour. Based on the pattern of carbon dioxide and ethylene production, pomegranate is classified as a nonclimacteric fruit (one that exhibits no dramatic changes in postharvest physiology or composition).[7,18]

14.6.2 EFFECTS OF TEMPERATURE AND RELATIVE HUMIDITY
ON DETERIORATION RATE

Both respiration and ethylene production rates increased with temperature. The Q_{10} values for respiration were 3.4 between 0°C and 10°C, 3.0 between 10°C and 20°C, and 2.3 between 20°C and 30°C.[7] Storage at 5°C or lower resulted in chilling injury; the severity increased with time and with lowered temperature. Chilling injury symptoms, which became more visible after transfer to 20°C for 3 days, included brown discoloration of the white locular septa separating the arils. Pomegranate can be stored at 5°C for up to 2 months, but longer storage should be at 7°C to avoid chilling injury.[7]

Pomegranates are very susceptible to water loss resulting in shriveling of the rind. The higher the temperature and the lower the relative humidity, the greater the water loss. Ideally, pomegranates should be kept at 90 to 95% relative humidity.[7,20] Use of plastic liners and waxing can reduce water loss, especially under conditions of lower relative humidity.[22,23]

14.6.3 RESPONSES TO ETHYLENE

Exposure of pomegranates to 1, 10, or 100 ppm ethylene in air for up to 13 days at 20°C stimulated their respiration rate in proportion to the ethylene concentration.[7] Subjecting pomegranates to 100 ppm ethylene in air for 2 days temporarily increased their respiration and ethylene production rates, which then declined to near the levels of control fruits after 3 days in air. This response occurred again when the fruits were exposed to a second 2-day ethylene treatment after 7 days in storage.[7] These responses are typical of nonclimacteric fruits.

None of the ethylene treatments had a significant effect on skin color, juice color, soluble solids, pH, or titratable acidity of the pomegranates. These results indicate that pomegranates do not ripen once removed from the tree and should be picked when fully ripe to ensure the best eating quality for the consumer. Also, there is no value in treating harvested pomegranates with ethylene.[7]

14.6.4 RESPONSES TO MODIFIED ATMOSPHERES

We evaluated the efficacy of atmospheric modification in controlling decay and maintaining quality of "Wonderful" pomegranates kept at 5°C, 7.5°C, or 10°C during one season using air, 2% oxygen, air + 10% carbon dioxide, and 2% oxygen + 10%

carbon dioxide. During another season we tested the following atmospheres at 5°C and 7.5°C: air, 5% oxygen, air + 10% carbon dioxide, 5% oxygen + 10% carbon dioxide, air + 15% carbon dioxide, 5% oxygen + 15% carbon dioxide. We found that it is possible to store pomegranates at 7.5°C in 5% oxygen + 15% carbon dioxide for up to 5 months, provided that the level of latent fungal infections at the time of harvest is low and that the pomegranates are sorted carefully after harvest to store only fruits that are free from defects and decay. Carbon dioxide-enriched atmospheres resulted in a lower synthesis rate of anthocyanins and other phenolic compounds[14] and higher concentrations of acetaldehyde, ethanol, and ethyl acetate, especially after 4 and 5 months of storage.[24] Accumulation of these volatiles was greater at 7.5°C than at 5°C, but in both cases the highest concentrations were below the threshold values for detection of off-flavors.[24]

Modified atmosphere packaging with appropriate polymeric films can be used to create a beneficial atmosphere (of 5–10% oxygen plus 10–15% carbon dioxide) during transport and storage of pomegranates.[22]

Scald is a physiological disorder that limits long-term storage of pomegranate fruits.[25,26] Scald symptoms developed mainly on the stem end of the fruit as brown discoloration on up to 60% of the skin without affecting the internal tissues. Among treatments tested, storage in a controlled atmosphere (CA) was the only treatment that successfully controlled this disorder.[26] The best CA combination was 5% O_2 + 15% CO_2, which resulted in a lower accumulation of fermentative metabolites than the other CA treatments.

14.6.5 PHYSIOLOGICAL DISORDERS

Appearance, especially red color, is an important quality factor for marketing fresh pomegranates. Many factors affect appearance, including bruising, water loss, decay, and the development of physiological disorders during storage. In general, the major cause limiting the storage potential of pomegranates is the development of decay, which is often caused by the presence of fungal inoculum in the blossom end of the fruit. This problem is aggravated at temperatures higher than 5°C, which are recommended for pomegranates to avoid chilling injury (internal tissue browning). For long-term storage, scald of the husk surface is another factor limiting storage life.

14.6.5.1 Chilling Injury

Pomegranates are susceptible to chilling injury if stored longer than 1 month at temperatures between their freezing point (–3°C) and 5°C, or longer than 2 months at 5°C.[7,18] Upon transfer to 20°C (simulated marketing conditions), respiration and ethylene production rates increase and other chilling injury symptoms (brown discoloration of the white locular septa and pale color of the arils) appear; their severity increases with lower temperatures and longer durations. Another consequence of chilling injury is increased susceptibility to decay.

The minimum safe temperature for postharvest handling of pomegranates ranges between 5°C and 8°C, depending on variety and production area.[7,18,20,27,28] Some studies have shown a reduction in incidence and severity of chilling injury symptoms by conditioning before storage, intermittent warming during storage, or modified atmosphere packaging.

14.6.5.2 Husk Scald

Scald symptoms appear as a superficial (skin) browning, similar to superficial scald of apples, which generally develops from the stem end of the fruit and then spreads toward the blossom end as it increases in severity.[25,26] Moreover, husk scald increases the susceptibility of the fruit to decay.

Scald incidence and severity were greater on pomegranates harvested during late season than on those harvested during midseason, indicating that this disorder may be associated with senescence. All pomegranates from both harvests that were kept in air exhibited some scald after 4 to 6 months at 7°C. Neither diphenylamine (DPA) nor 1-methylcyclopropene (1-MCP) alone or together reduced scald incidence and severity.[26] In contrast, the three controlled atmosphere (CA) storage conditions tested significantly reduced scald incidence and severity on pomegranates from both harvest dates for up to 6 months at 7°C.[26] However, the two CA treatments with 1% O_2 resulted in greater accumulation of fermentative volatiles (acetaldehyde, ethanol, and ethyl acetate) than the CA treatment with 5% O_2, especially in the midseason-harvested pomegranates. In addition to its fungistatic effects, 15% CO_2 appears to be critical for inhibition of scald development on pomegranates. These results confirm our recommendation of 5% O_2 + 15% CO_2 (balance N_2) as the optimal CA for pomegranates at 7°C and 90 to 95% relative humidity.[24] CA storage (5% O_2 + 15% CO_2) also decreased or prevented changes in carotenoid, acyl lipid, and phenylpropanoid metabolism that were associated with scald development in stem-end peel tissue of air-stored fruit.[26]

14.7 POSTHARVEST PATHOLOGY

14.7.1 Decay-Causing Pathogens

Heart rot may be caused by *Aspergillus* spp. or *Alternaria* spp. Infection begins in the orchard, especially following rain during flowering and early fruit development. The fungi can grow within the fruit without external symptoms except for slightly abnormal skin color. If the mass of blackened arils reaches the rind, it will cause softening of the affected area; these pomegranates can be detected and removed by the sorters in the packinghouse.[2]

Gray mold caused by *Botrytis cinerea* is the most important postharvest decay-causing fungus. Infection can begin in the orchard and the spores can be in the calyx area at harvest time. As it grows, the skin turns light brown, tough, and leathery followed by appearance of gray mycelial growth.

14.7.2 Decay Control Methods

Minimizing physical damage during harvesting and postharvest handling plus maintaining optimal temperature and relative humidity throughout postharvest handling of pomegranates are very important decay control strategies. Carbon-dioxide-enriched atmospheres are fungistatic and inhibit growth of *Botrytis cinerea*. Use of Fludioxonil (Scholar) as a postharvest fungicide is effective in controlling this fungus and is approved by USEPA with a maximum residue limit of 5 ppm.

14.8 POSTHARVEST HANDLING SYSTEMS

14.8.1 PREPARATION FOR MARKET

Pickers harvest pomegranates with clippers and place the fruits in picking bags for transfer to harvest bins that will be transported to the packinghouse. Then the pomegranates are sorted to eliminate those with severe defects (such as scuffing, cuts, bruises, splitting, and decay) and the remaining fruits are separated according to the magnitude of the physical defects. Pomegranates with moderate defects are used for processing into juice and those with slight or no defects are marketed fresh. The latter fruits are washed, air dried to remove surface moisture, fungicide treated, waxed, divided into several size categories, and packed in shipping containers. Various ways to immobilize the fruits within the shipping containers may be used to reduce incidence and severity of scuffing and impact bruising during handling. Perforated plastic box liners may be used to reduce water loss during postharvest handling of pomegranates. Packed fruits are cooled by forced-air cooling to 7°C and kept at that temperature and 90 to 95% relative humidity during storage and transport to retail distribution centers.

14.8.2 OPTIMAL STORAGE CONDITIONS

Several postharvest conditions have been evaluated for long-term storage of pomegranates, including low temperature, delayed harvest, intermittent warming,[29,30] controlled atmosphere (CA),[22,24,31,32] and partial drying.[33] Among these procedures, the most successful in reducing decay and physiological disorders is the use of CA storage, which, with a combination of 5% O_2 and 15% CO_2, has been shown to extend pomegranate postharvest life for up to 5 months at 7°C. This combination also avoids the accumulation of high levels of ethanol, observed under CA conditions with lower levels of oxygen, which limits the marketability of the fruit.

Optimal storage temperature ranges from 5°C to 8°C, depending on the variety and production area; 7°C is recommended for "Wonderful" pomegranates. In all cases, 90 to 95% relative humidity should be maintained in the surrounding atmosphere. Storage potential ranges from 3–4 months in air and from 4–6 months in a controlled atmosphere of 5% oxygen + 15% carbon dioxide (balance nitrogen). Pomegranates should not be mixed with grapes, since they can be damaged from exposure to the sulfur dioxide that is used to control gray mold on grapes.

14.9 FACTORS AFFECTING QUALITY AND SAFETY OF POMEGRANATE ARILS

One of the limiting factors to increased consumption of fresh pomegranates is the effort needed to extract the arils from the fruit. Thus, providing the consumer with value-added, ready-to-eat pomegranate arils may help increase consumption. Gil et al.[34,35] and Hess-Pierce and Kader[36,37] investigated the effects of preextraction storage duration and postextraction packaging and handling conditions on deterioration rate of pomegranate arils (juice sacs surrounding the seeds). Pomegranate arils have relatively low rates of respiration (1.5–3 and 3–6 ml carbon dioxide per kilogram per hour at 5°C and 7°C, respectively) and ethylene production (5–15 and 15–30 nl ethylene per kilogram per hour at 5°C and 7°C, respectively). It is possible to produce arils that retain good sensory and microbial quality for up to 14 days of shelf life at 5°C from pomegranate fruits that are stored at 7°C for up to 3 months in air or up to 5 months in a controlled atmosphere of 5% oxygen + 15% carbon dioxide + 85% nitrogen. Mechanical damage to the arils must be minimized during their extraction from the fruit, washing, drying to remove surface moisture, and packaging, since damaged arils are more susceptible to decay-causing fungi. Carbon-dioxide-enriched atmospheres have a fungistatic effect and their optimal range for decay control without inducing off-flavors in the arils is 15 to 20% carbon dioxide added to either air or 5% oxygen. Although intact pomegranate fruits are chilling-sensitive, the arils are chilling-tolerant and should be kept at temperatures between 0°C and 5°C to maintain their quality and microbial safety. Pomegranate arils that are not damaged or microbially contaminated can be kept at 0°C for up to 21 days, at 2°C for up to 18 days, or at 5°C for up to 14 days in marketable condition.[37]

REFERENCES

1. Hodgson, R.W., The pomegranate, *Bulletin No. 276*, University of California Press, Berkeley, 1917.
2. LaRue, J.H., Growing pomegranates in California, University of California Division of Agricultural Sciences Leaflet 2459, 1980.
3. Adsule, R.N. and Patil, N.B., Pomegranate, in *Handbook of fruit science and technology,* Salunkhe, D.K. and Kadam, S.S., eds., Marcel Dekker, New York, 1995, 455.
4. Kumar, G.N.M., Pomegranate, in *Fruits of tropical and subtropical origin,* Nagy, S. et al., eds., Florida Science Source, Lake Alfred, FL, 1990, 328.
5. Patil, A.V. and Karale, A.R., Pomegranate, in *Fruits: Tropical and subtropical,* Kose, T.K. and Mitra, S.K., eds., Naya Prokash, Calcutta, India, 614.
6. Roy, S.K. and Waskar, D.P., Pomegranate, in *Postharvest physiology and storage of tropical and subtropical fruits,* CAB International, Wallingford, UK, 1997, 365.
7. Kader, A.A., Chordas, A., and Elyatem, S.M., Responses of pomegranates to ethylene treatment and storage temperature, *Calif. Agric.,* 38, 7&8 , 14, 1984.
8. Gil, M.I., Sanchez, R., Marin, J.G., and Artes, F., Quality changes in pomegranates during ripening and cold storage, *Z. Lebensm. Unters. Forsch.,* 202, 481, 1996.
9. Al-Maiman, S.A. and Ahmad, D., Changes in physical and chemical properties during pomegranate (*Punica granatum* L.) fruit maturation, *Food Chem.,* 76, 437, 2002.
10. Lee, S.W., Kim, K.S., and Kim, S.D., Studies on the compositional changes of pomegranate fruit during maturation. I. Changes in sugars, organic acids, amino acids, and the respiration rate, *J. Korean Soc. Hort. Sci.,* 15, 57, 1974.
11. Du, C.T., Wang, P.L., and Francis, F.J., Anthocyanins of pomegranates, *Punica granatum, J. Food Sci.,* 40, 417, 1975.

12. Gil, M.I. et al., Influence of cultivar, maturity stage, and geographical location on the juice pigmentation of Tunisian pomegranates, *Z. Lebensm. Unters. Forsch.,* 201, 361, 1995.

13. Gil, M.I. et al., Changes in pomegranate juice pigmentation during ripening, *J .Food Sci.,* 68, 77, 1995.

14. Holcroft, D.M., Gil, M.I., and Kader, A.A., Effect of carbon dioxide on anthocyanins, phenylalanine ammonia lyase and glucosyltransferase in the arils of stored pomegranates, *J. Amer. Soc. Hort. Sci.,*123, 136, 1998.

15. Lee, S.W., Kim, K.S., and Kim, S.D., Studies on the compositional changes of pomegranate fruit during maturation. II. Changes in polyphenol compounds and anthocyanin pigments, *J. Korean Soc. Hort. Sci.,* 15, 64, 1974.

16. Gil, M.I. et al., Antioxidant activity of pomegranate juice and its relationship with phenolic composition and processing, *J. Agric. Food Chem.,* 48, 4581, 2000.

17. Li, Y. et al., Evaluation of antioxidant properties of pomegranate peel extract in comparison with pomegranate pulp extract, *Food Chem.,* 96, 254, 2006.

18. Elyatem, S.M. and Kader, A.A., Postharvest physiology and storage behaviour of pomegranate fruits, *Scientia Hort.,* 24, 287, 1984.

19. Ben-Arie, R., Segal, N., and Guelfat-Reich, The maturation and ripening of the 'Wonderful' pomegranate, *J. Amer. Soc. Hort. Sci.,* 109, 898, 1984.

20. Crisosto, C.H., Mitcham, E.J., and Kader, A.A., Pomegranates, in *Produce Facts,* available online at http://postharvest.ucdavis.edu/Produce/ProduceFacts/Fruit/Pomegranate.shtml

21. Melgarejo, P. et al., Kaolin treatment to reduce pomegranate sunburn, *Scientia Hort.,* 100, 349, 2004.

22. Artes, F., Villaescusa,R., and Tudela, J.A., Modified atmosphere packaging of pomegranate, *J. Food Sci.,* 65, 1112, 2000.

23. Nanda, S., Sudhakar Rao, D.V., and Krishnamurthy, S., Effects of shrink film wrapping and storage temperature on the shelf life and quality of pomegranate fruits cv. Ganesh, *Postharv. Biol. Technol.,* 22, 61, 2001.

24. Hess-Pierce, B.M. and Kader, A.A., Responses of 'Wonderful' pomegranates to controlled atmospheres, *Acta Hort.,* 600, 751, 2003.

25. Ben-Arie, R. and Or, E., The development and control of husk scald on 'Wonderful' pomegranate fruit during storage, *J. Amer. Soc. Hort. Sci.,* 111, 395, 1986.

26. Defilippi, B.G., Whitaker, B.D., Hess-Pierce, B.M., and Kader, A.A., Development and control of scald on Wonderful pomegranates during long-term storage, *Postharv. Biol. Technol.,* in press, 2006.

27. Koksal, I., Research on the storage of pomegranates cv. 'Gok Bahce' under different conditions, *Acta Hort.,* 258, 295, 1989.

28. Pekmezci, M. and Erkan, M., Pomegranate, in *USDA Agricultural Handbook 66,* available online at http://www.ba.ars.usda.gov/hb66/index.html.

29. Artes, F., Tudela, J.A., and Gil, M.I., Improving the keeping quality of pomegranate fruit by intermittent warming, *Z. Lebensm. Unters. Forsch.,* 207, 316, 1998.

30. Artes, F., Tudela, J.A., and Villaescusa,R., Thermal Postharvest treatments for improving pomegranate quality and shelf life, *Postharv. Biol. Technol.,* 18, 245, 2000.

31. Artes, F., Marin, J.G., and Martinez, J.A., Controlled atmosphere storage of pomegranate, *Z. Lebensm. Unters. Forsch.,* 203, 33, 1996.

32. Kupper, W., Pekmezci, M., and Henze, J., Studies on CA-storage of pomegranate (*Punica granatum* L., cv. Hicaz), *Acta Hort.,* 398, 101, 1995.

33. Al-Kahtani, H.A., Intercultivar differences in quality and postharvest life of pomegranates influenced by partial drying, *J. Amer. Soc. Hort. Sci.,* 117, 100, 1992.

34. Gil, M.I., Artes, F., and Tomas-Barberan, F.A., Minimal processing and modified atmosphere packaging effects on pigmentation of pomegranate seeds, *J. Food Sci.,* 61, 161, 1996.

35. Gil, M.I., Martinez, and Artes, F., Minimally processed pomegranate seeds, *Lebensmittel-Wissenchaft Technologie,* 29, 708, 1996.

36. Hess-Pierce, B.M. and Kader, A.A., Carbon dioxide-enriched atmospheres extend postharvest life of pomegranate arils, in *CA-97 Program and Abstracts,* University of California, Davis, Calif., U.S., 1997, 135.

37. Hess-Pierce, B.M. and Kader, A.A., Factors affecting shelf-life of pomegranate arils, presented at the International Fresh-cut Produce Association 17th Annual Conference, Reno, Nevada, April 22–24, 2004.

Section 5

Summary

15 Summary

Risa N. Schulman

CONTENTS

The field of pomegranate research has experienced tremendous growth in the last eight to ten years. It started with a very small group of researchers, but their exciting preliminary results attracted additional scientists with varied backgrounds. Surprisingly at first, they started to see effects in a number of different disease areas, with multiple, sometimes overlapping, mechanisms of action. Evidence grew with further research, showing that beyond its potent effects in several diseases, the pomegranate is extraordinary because of its multifunctionality. This complements a trend in our understanding of illness today — that pathologies are intertwined with one another. Medical science has discovered links between obesity and diabetes, heart disease and erectile dysfunction, poor diet and cancer. That the pomegranate can impact illness in many ways makes it a highly valuable tool. Its multifunctionality also makes it an ideal research pathway for understanding the common underlying mechanisms of the major diseases facing us today, and how phytochemicals can intervene. The pomegranate is truly a twenty-first-century natural medicine.

The last five years have seen an explosion of published work on pomegranates, from fewer than 30 papers published before 2000, to more than 40 published in 2005 alone (Figure 15.1). The many outstanding scientists in this field have remained cautious, objective, and rigorous with their studies and have laid an excellent and rapidly expanding foundation for future work. As the field is poised to become a significant international effort, it was our intention to capture the current state of research so that all involved can avail themselves of what is known and where there is to go. Even as we write, new areas of inquiry are coming to light in brain health,[1,2] erectile dysfunction,[3] and diabetes.[4] It is critical that information exchange continue between researchers investigating these areas; it is in these intersections that the most

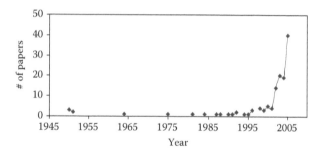

FIGURE 15.1 Journal articles on pomegranate, 1945 to 2005. Data derived from Medline.

exciting discoveries lie. The message of medicinal plants serves us well — they perform best when all the actives work together, rather than each as an isolated ingredient.

Each chapter of this book brought to light areas for continued exploration. We compile them here, as fruit for the imagination.

15.1 BIOCHEMISTRY

- Identification and characterization (chemical and biological) of all actives (Chapter 1)
- Interaction of pomegranate actives with intestinal receptors vs. interaction with bodily receptors after metabolism (Chapter 1)
- Development of novel methods of chromatographic separation to improve our understanding of the composition of pomegranate phytochemicals (Chapter 1)
- Investigation of the effects of specific pomegranate actives, their biological fate and effects *in vivo,* and their mechanisms of action (Chapter 1)
- Evaluation of the possible role of ellagitannins and ellagic acid as dietary "prophytoestrogens" (Chapter 3)
- Evaluation of the activity of pomegranate metabolites at concentrations reached in plasma and tissues (Chapter 3)
- Nutrient–gene interactions
- DNA microarray studies

15.2 CARDIOVASCULAR

- Clinical and nutritional studies using combinations of several types of antioxidants, and using reliable biological markers of oxidative stress clinical and nutritional studies done in populations suitable for antioxidant treatment (Chapter 4)
- Interaction of pomegranate actives with angiotensin II (Chapter 4)

15.3 CANCER

- Understanding the mechanisms and active ingredients behind anticancer effects (Chapter 6)
- Clinical trials in human cancers (Chapter 8)

- Development of dietary agent or botanical drug containing a mixture of actives (Chapter 6)
- Conducting in-depth molecular studies and elucidate the components responsible for beneficial effects in skin cancer (Chapter 7)
- For skin cancer applications, determining optimal dosing, period, and route of administration and toxicity, and assessing the bioavailability of the active principle for animal and human studies, if warranted (Chapter 7)

15.4 ESTROGEN STUDIES

- Continued investigation of whether the pomegranate has estrogenic effects after long-term use (Chapters 9, 10)

15.5 ANTIMICROBIAL

- Investigation of compound-specific antimicrobial activity (Chapter 11)
- Studying synergistic effects of pomegranate extracts with antibiotics (Chapter 11)
- Investigation of the efficacy of pomegranate in food preservation (Chapter 11)

15.6 COMMERCIALIZATION

- Development of well-characterized and standardized pomegranate botanical extracts for human and animal studies (Chapter 12)
- Use of the spent by-product material generated from the commercial juice industry to generate ellagitannin-rich extracts for use as botanical ingredients (Chapter 12)

15.7 PLANT GROWTH AND IMPROVEMENT

- Take advantage of intact genetic diversity to develop elite pomegranate varieties (Chapter 13)

CONCLUSION

It is our hope that this book accomplishes its task, and that the beauty and intrigue of the pomegranate will continue to captivate and inspire us to unlock its inner mysteries and give them to humankind as a healing gift. I leave it to the poet to say it best:

"Pomegranate," which if cut deep down the middle
Shows a heart within blood-tinctured, of a veined humanity.

Elizabeth Barrett Browning

REFERENCES

1. Loren, D.J., Seeram, N.P., Schulman, R.N., and Holtzman, D.M., Maternal dietary supplementation with pomegranate juice is neuroprotective in an animal model of neonatal hypoxic-ischemic brain injury, *Pediatr. Res.* 57(6):858–64, 2005.
2. Sweeney, M.I., Dietary bioactive compounds, modulation of physiological processes I. Presented at FASEB, April 2–6, 2005, San Diego, CA.
3. Azadzoi, K.M., Schulman, R.N., Aviram, M., and Siroky, M.B., Oxidative stress in arteriogenic erectile dysfunction: prophylactic role of antioxidants, *J. Urol.* 174(1):386–93, 2005.
4. Rozenberg, O., Howell, A. and Aviram, M., Pomegranate juice sugar fraction reduces macrophage oxidative state, whereas white grape juice sugar fraction increases it, *Atherosclerosis,* in press, 2005.

Index

References to figures and tables are given in italics.